National Fire Service
Incident Management System

Model Procedures Guide for Structural Firefighting

First Edition

Prepared by the National Fire Service Incident Management System Consortium Model Procedures Committee

Published by

Fire Protection Publications
Oklahoma State University

The following organizations have endorsed this model procedure: Fireground Command • FIRESCOPE • International Association of Fire Chiefs • International Association of Fire Fighters • International Society of Fire Service Instructors • Society of National Fire Academy Instructors • International Fire Service Training Association

Many other fire service organizations are presently reviewing the National Fire Service Incident Management Systems Consortium's request for endorsement of the model procedure. As endorsements are received, they will be listed in future printings of the procedure.

The Consortium thanks those organizations endorsing the procedure.

ISBN 0-87939-108-1
Library of Congress 93-74000

Photos courtesy of:
Phoenix Fire Department, Oklahoma City Fire Department

FIRST PRINTING: 11/93, SECOND PRINTING 9/95

Statement of Purpose

This document is designed as a guide to assist emergency service organizations in the implementation of an Incident Management System. It may also serve as the initial design document from which your jurisdiction may operate during emergency incidents. Once adopted by the jurisdiction and with appropriate training of personnel, it provides an easily understood organizational structure and procedures to follow during emergency incidents.

This model encourages the use of an Incident Management System for structural fire incidents. It provides for its use on small, routine incidents as well as allowing for expansion of the organization to meet the needs of an escalating incident.

Please read this document often. The daily use of its concepts and procedures will encourage jurisdictionwide and statewide standardization to assist all emergency providers to become better organized, more adaptive and more expedient in their incident management service delivery.

Table of Contents

Foreword

National Fire Service Incident Management System Model for Structural Firefighting

The purpose of the Incident Management System is to provide for a systematic development of a complete, functional Command organization designed to allow for single or multi-agency use which increases the effectiveness of Command and firefighter safety.

This model system was developed by the National Fire Service Incident Management System Consortium. It combines Command strategy with organizational procedures and is designed to be used primarily for structural fire incidents using up to 25 fire companies. Much of the organizational design is applicable to other types of emergency incidents. The model reflects the merger of certain elements of the California FIRESCOPE Incident Command System and the Phoenix Fireground Command System.

The key elements of the system are:

- The systematic development of a complete, functional organization with the major functions being Command, operations, planning, logistics, and finance/administration.
- Designed to allow for multi-agency adoption in federal, state, and local fire agencies. Therefore, organizational terminology used in the Incident Management System is designed to be acceptable to all levels of government.
- Designed to be the basic, everyday operating system for all incidents within each agency. Therefore, the transition to large and/or multi-agency operations requires a minimum of adjustment for any of the agencies involved.
- The organization builds from the ground up, with the management of all major functions initially being the responsibility of one or just a few persons. Functional units are designed to handle the most important incident activities. As the incident grows in size and/or complexity, functional unit management is assigned to additional individuals in order to maintain a reasonable span of control and efficiency.
- Designed on the premise that the jurisdictional authority of the involved agencies will not be compromised. Each agency having legal responsibility within its jurisdiction is assumed to have full Command authority within its jurisdiction at all times. Assisting agencies will normally function under the direction of the Incident Commander appointed by the jurisdiction within which the incident occurs.
- Multi-jurisdictional incidents will normally be managed under a Unified Command management structure involving a single incident Command Post and a single Incident Action Plan—applicable to all agencies involved in the incident.

- The system is intended to be staffed and operated by qualified personnel from any agency, and a typical incident could involve the use of personnel from a variety of agencies, working in many different parts of the organization.
- The system expands and contracts organizationally based upon the needs of the incident. Span-of-control recommendations are followed closely; therefore, the organizational structure is never larger than required.

Although the focus of the Consortium's work is structural fire—the Consortium recognizes the importance to the fire service of coordinating incident response with responders of other disciplines, such as medical, law enforcement, and public works. An effective incident management system must provide an integrated multi-discipline approach. The Incident Management System model, while capitalizing on the strengths of fireground Command, provides an overall structure that allows the successful integration of multiple disciplines, allowing application to the "all risk" nature of emergency incidents.

Other response disciplines (law enforcement, public works) are encouraged to address their specific tactical needs within the Command/Operations Sections in the detail given to fireground Command, while retaining the overall Incident Management System structure. On multi-discipline incidents, experience has proven the critical necessity of integrating response agencies into one operational organization managed and supported by one structure. For this reason, the Consortium supports an integrated, multi-discipline organization over separate incident management systems for each organization.

The National Fire Service Incident Management System Consortium believes that any incident management system should be procedures-driven for the following reasons:

- Written procedures reflect department policy on incident management.
- Procedures provide a standardized approach to managing any incident.
- Procedures provide predictable approaches to incident management.
- Procedures should be applied routinely.
- Procedures provide a training tool for firefighters' reference.
- Procedures provide a baseline for critiques and review of incidents.
- Procedures make the Incident Commander's operations more effective.

This model reflects a procedural approach to the overall organization structure of the Incident Management System. The Consortium will be addressing various models of other "all risk" types of urban emergencies (such as multi-casualty, hazardous materials, highrise) in future work.

1

Command Procedures

1

Command Procedures

Purpose

Fire Departments respond to a wide range of emergency incidents. This procedure guide identifies standard operating procedures that can be employed in establishing Command. The system provides for the effective management of personnel and resources providing for the safety and welfare of personnel. It also establishes procedures for the implementation of all components of the Incident Management System for structural/fire operations.

Command Procedures are designed to:

- Fix the responsibility for Command on a specific individual through a standard identification system, depending on the arrival sequence of members, companies, and chief officers.
- Ensure that a strong, direct, and visible Command will be established from the onset of the incident.
- Establish an effective incident organization defining the activities and responsibilities assigned to the Incident Commander and to other individuals operating within the Incident Management System.
- Provide a system to process information to support incident management, planning, and decision making.
- Provide a system for the orderly transfer of Command to subsequent arriving officers.

Responsibilities of Command

The Incident Commander is responsible for the completion of the tactical priorities. **The Tactical Priorities** are:

1. Remove endangered occupants and treat the injured.
2. Stabilize the incident and provide for life safety.
3. Conserve property.
4. Provide for the safety, accountability, and welfare of personnel. **This priority is ongoing throughout the incident.**

The Incident Management System is used to facilitate the completion of the tactical priorities. The **Incident Commander** is the person who drives the Incident Management System towards that end. The Incident Commander is responsible for building a Command structure that matches the organizational needs of the incident to achieve the completion of the tactical priorities for the incident. The Functions of Command define standard activities that are performed by the Incident Commander to achieve the Tactical Priorities.

Functions of Command

The Functions of Command Include:

- Assume and announce Command and establish an effective operating position (Command Post).
- Rapidly evaluate the situation (size up).
- Initiate, maintain, and control the communications process.
- Identify the overall strategy, develop an incident action plan, and assign companies and personnel consistent with plans and standard operating procedures.
- Develop an effective Incident Management Organization.
- Provide tactical objectives.
- Review, evaluate, and revise (as needed) the incident action plan.
- Provide for the continuity, transfer, and termination of Command.

The Incident Commander is responsible for all of these functions. As Command is transferred, so is the responsibility for these functions. The first five (5) functions must be addressed immediately from the initial assumption of Command.

Establishing Command

The first fire department member or unit to arrive at the scene shall assume Command of the incident. The initial Incident Commander shall remain in Command until Command is transferred or the incident is stabilized and terminated.

1. The first unit or member on the scene must initiate whatever parts of the Incident Management System are needed to effectively manage the incident scene.
2. A single company incident (trash fires, single patient E.M.S. incidents, etc.) may only require that Company or unit acknowledge their arrival on the scene.

3. For incidents that require the commitment of multiple Companies or units, the first unit or member on the scene must establish and announce "Command," and develop an Incident Command Structure appropriate for the incident.

The first arriving fire department unit activates the Command process by giving an initial radio report.

The Radio Report should include:

1. Unit designation of the unit arriving on the scene.
2. A brief description of the incident situation, (i.e., building size, occupancy, Haz Mat release, multi-vehicle accident, etc.)
3. Obvious conditions (working fire, Haz Mat spill, multiple patients, etc.).
4. Brief description of action taken.
5. Declaration of Strategy (this applies to structure fires).
6. Any obvious safety concerns.
7. Assumption, identification, and location of Command.
8. Request or release resources as required.

Example:

For an offensive structure fire–

"Engine Eleven is on the scene of a large two story school with a working fire on the second floor. Engine Eleven is laying a supply line and going in with a handline to the second floor for search and rescue. This is an offensive fire attack. Engine Eleven will be 7th Street Command."

For a defensive structure fire–

"Engine One is on the scene of a medium size warehouse fully involved with exposures to the east. Engine One is laying a supply line and attacking the fire with a master stream and handline to the exposure for search and rescue and fire attack. This is a defensive fire. Engine One will be Buckeye Command."

For an E.M.S. incident–

"Ladder 11 is on the scene with a multi-vehicle accident. Give me the balance of a 1st Alarm medical with three ambulances. Ladder 11 will be Parkway Command."

For a single Company Incident–

"Engine 6 is on the scene of a dumpster fire with no exposures. Engine 6 can handle."

Radio Designation:

The radio designation "Command" will be used along with the geographical location of the incident (i.e., "7th Street Command," "Metro Center Command"). This designation will not change throughout the duration of the incident. The designation of "Command" will remain with the officer currently in Command of the incident throughout the event.

Command Options

The responsibility of the first arriving unit or member to assume Command of the incident presents several options, depending on the situation. If a Chief Officer, member, or unit without tactical capabilities (i.e., staff vehicle, no equipment, etc.) initiates Command, the establishment of a Command Post should be a top priority. At most incidents the initial Incident Commander will be a Company Officer. The following Command options define the Company Officer's direct involvement in tactical activities and the modes of Command that may be utilized.

Nothing Showing Mode:

These situations generally require investigation by the initial arriving company while other units remain in a staged mode. The officer should go with the company to investigate while utilizing a portable radio to Command the incident.

Fast Attack Mode:

Situations that require immediate action to stabilize and require the Company Officer's assistance and direct involvement in the attack. In these situations the Company Officer goes with the crew to provide the appropriate level of supervision. Examples of these situations include:

- Offensive fire attacks (especially in marginal situations).
- Critical life safety situations (i.e., rescue) which must be achieved in a compressed time.
- Any incident where the safety and welfare of firefighters is a major concern.
- Obvious working incidents that require further investigation by the Company Officer.

Where fast intervention is critical, utilization of the portable radio will permit the Company Officer's involvement in the attack without neglecting Command responsibilities. The Fast Attack mode should not last more than a few minutes and will end with one of the following:

A. The situation is stabilized.

B. The situation is not stabilized and the Company Officer must withdraw to the exterior and establish a Command Post. At some time the Company Officer must decide whether or not to withdraw the remainder of the crew, based on the crew's capabilities and experience, safety issues, and the ability to communicate with the crew. No crew should remain in a hazardous area without radio communications capabilities.

C. Command is transferred to another Higher Ranking Officer. When a Chief Officer is assuming Command, the Chief Officer may opt to return the Company Officer to his/her crew, or assign him/her to a subordinate position.

Command Mode:

Certain incidents, by virtue of their size, complexity, or potential for rapid expansion, require immediate strong, direct, overall Command. In such cases, the Company Officer will initially assume an exterior, safe, and effective Command position and maintain that position until relieved by a Higher Ranking Officer. **A tactical worksheet shall be initiated and utilized to assist in managing this type of incident (See Appendix C).**

If the Company Officer selects the Command mode, the following options are available regarding the assignment of the remaining crew members.

A. The officer may "move up" within the company and place the company into action with the remaining members. One of the crew members will serve as the acting Company Officer and **should be provided with a portable radio**. The collective and individual capabilities and experience of the crew will regulate this action.

B. The officer may assign the crew members to work under the supervision of another Company Officer. In such cases, the Officer assuming Command must communicate with the Officer of the other company and indicate the assignment of those personnel.

C. The officer may elect to assign the crew members to perform staff functions to assist Command.

A Company Officer assuming Command has a choice of modes and degrees of personal involvement in the tactical activities, but continues to be fully responsible for the Command functions. The initiative and judgment of the Officer are of great importance. The modes identified are guidelines to assist the Officer in planning appropriate actions. The actions initiated should conform with one of the above mentioned modes of operation.

Passing Command

In certain situations, it may be advantageous for a first arriving Company Officer to pass Command to the next company **on the scene**. This is indicated when the initial commitment of the first arriving company requires a full crew (i.e., high-rise or an immediate rescue situation) and another company is on the scene.

"Passing Command" to a unit that is not on the scene creates a gap in the Command process and compromises incident management. To prevent this "gap," **Command shall not be passed to an officer who is not on the scene.** It is preferable to have the initial arriving Company Officer continue to operate in the fast attack mode until Command can be passed to an on-scene unit.

When a Chief Officer arrives at the scene at the same time as the initial arriving company, the Chief Officer should assume Command of the incident.

Should a situation occur where a later arriving Company or Chief Officer cannot locate or communicate with Command (after several radio attempts), they will assume and announce their assumption of Command and initiate whatever actions are necessary to confirm the safety of the missing crew.

Transfer of Command

Command is transferred to improve the quality of the Command organization. The following guidelines outline the transfer of Command process. The transfer of Command through various ranking officers must be predetermined by the local departments. Below is an example.

1. The first fire department member arriving on the scene will automatically assume Command. This will normally be a Company Officer, but could be any fire department member up to and including the Fire Chief.
2. The first arriving Company Officer will assume Command after the transfer of Command procedures have been completed (assuming an equal or higher ranking officer has not already assumed Command).

3. The first arriving Chief Officer should assume Command of the incident following transfer of Command procedures.

4. The second arriving Chief Officer should report to the Command Post for assignment.

5. Later arriving, higher-ranking Chief Officers may choose to assume Command, or assume adviser positions.

6. Assumption of Command is discretionary for Assistant Chiefs and the Fire Chief.

Within the chain of Command, the actual transfer of Command will be regulated by the following procedure:

1. The Officer assuming Command will communicate with the person being relieved by radio or face-to-face. Face-to-face is the preferred method to transfer Command.

2. The person being relieved will brief the officer assuming Command indicating at least the following:

    A. Incident conditions (fire location and extent, Haz Mat spill or release, number of patients, etc.)

    B. Incident action plan.

    C. Progress toward completion of the tactical objectives.

    D. Safety considerations.

    E. Deployment and assignment of operating companies and personnel.

    F. Appraisal of need for additional resources.

3. The person being relieved of Command should review the tactical worksheet with the Officer assuming Command. This sheet provides the most effective framework for Command transfer as it outlines the location and status of personnel and resources in a standard form that should be well known to all members.

The person being relieved of Command will be assigned to best advantage by the Officer assuming Command.

General Considerations

The response and arrival of additional ranking officers on the incident scene strengthens the overall Command function. As the incident escalates, the Incident Commander should use these Subordinate Officers as needed.

A fire department's communications procedures should include communications necessary to gather and analyze information to plan, issue orders, and supervise operations.

For example:

- Assignment completed.
- Additional resources required.
- Unable to complete.
- Special information.

The arrival of a ranking officer on the incident scene does not mean that Command has been transferred to that officer. Command is only transferred when the outlined transfer-of Command process has been completed.

Chief Officers and Staff Personnel should report directly to a designated location for assignment by the Incident Commander.

The Incident Commander has the overall responsibility for managing an incident. Simply stated the Incident Commander has complete authority and responsibility for the Incident. * If a higher ranking officer wants to effect a change in the management of an incident, they must first be on the scene of the incident, then utilize the transfer-of-Command procedure.

* Anyone can effect a change in incident management in extreme situations relating to safety by notifying Command and initiating corrective action.

2

Command Structure

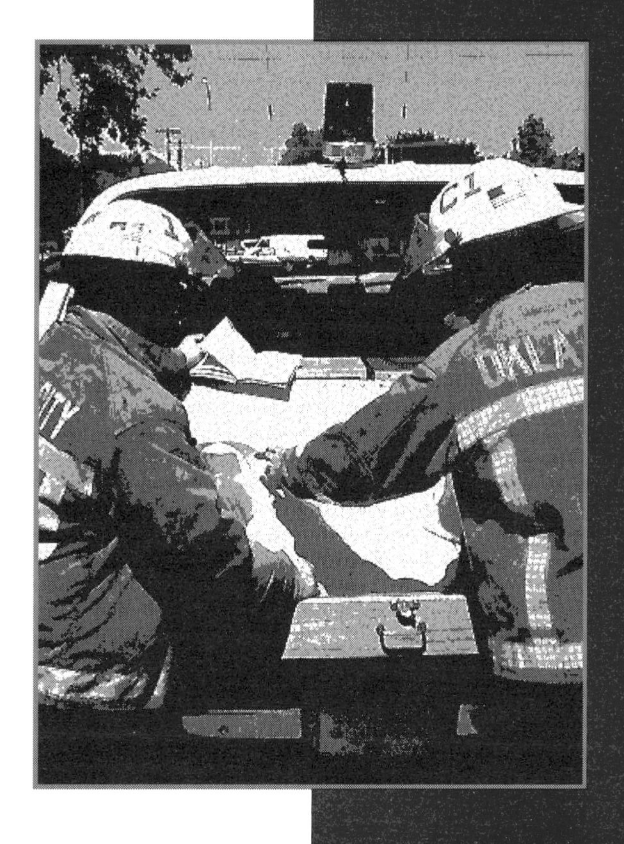

2
Command Structure

It will be the responsibility of the Incident Commander to develop an organizational structure utilizing standard operating procedures as soon as possible after arrival and implementation of initial tactical control measures. The size and complexity of the organizational structure, obviously, will be determined by the scope of the emergency.

Incident Management System Operations

The Incident Management System should be considered the basic incident management system to be used on any size or kind of incident. The only change in using the Incident Management System on a very large incident rather than a small incident is the method of growth of the basic emergency management organization to meet the increased needs. Thus, the full establishment of the Incident Management System should be viewed as an extension of the existing incident organization. The determination to expand the organization will be that of Command and would be done when a determination is made that the initial attack or reinforced attack will be insufficient. That determination would be made by the Incident Commander at the scene.

IMS Organizational Development

The following examples are guides in using the basic Incident Management System Organization for various size incidents.

Initial Response	1-5 Increments/1st Alarm
Reinforced Response	Greater Alarm/Mutual Aid

Initial Response

The first arriving unit or officer will assume Command until arrival of a higher ranking officer.

Upon arrival of a higher ranking officer, they will be briefed by the on-scene Incident Commander. The higher ranking officer will then assume Command. This transfer of Command is to be announced. The officer being relieved of Command responsibilities will be reassigned by the new Incident Commander.

Reinforced Response

A reinforced response will be initiated when the on-scene Incident Commander determines that the initial response resources will be insufficient to deal with the size or complexity of the incident.

Command Organization

The Command organization must develop at a pace which stays ahead of the tactical deployment of personnel and resources. In order for the Incident Commander to manage the incident, they must first be able to direct, control, and track the position and function of all operating companies. Building a Command organization is the best support mechanism the Incident Commander can utilize to achieve the harmonious balance between managing personnel and incident needs. Simply put, this means:

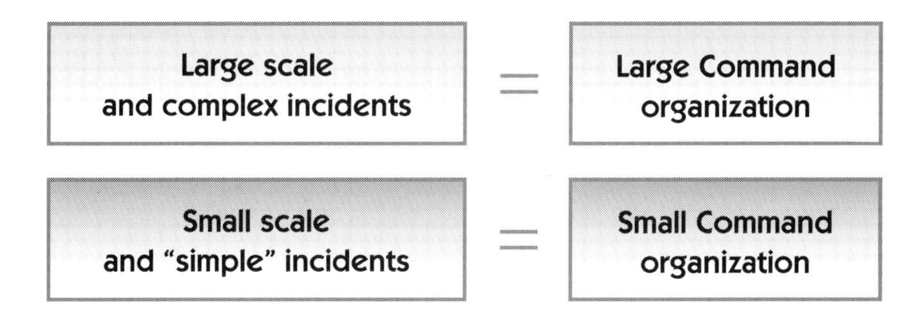

Note: The Incident Commander should have more people working than Commanding.

The basic configuration of Command includes three levels:

Strategic level:
overall direction of the incident

Tactical level:
assigns operational objectives

Task level:
specific tasks assigned to companies

Strategic Level:

The Strategic level involves the overall Command of the incident. The Incident Commander is responsible for the strategic level of the Command structure. The action plan should cover all strategic responsibilities, all tactical objectives, and all support activities needed during the entire operational period. The Action Plan defines where and when resources will be assigned to the incident to control the situation. This plan is the basis for developing a Command organization, assigning all resources, and establishing tactical objectives.

The strategic level responsibilities include:

Offensive or **Defensive** (These should be well defined in S.O.P.'s)

- Determining the appropriate strategy
- Establish overall incident objectives.
- Setting priorities.
- Develop an incident action plan.
- Obtaining and assigning resources.
- Predicting outcomes and planning.
- Assigning specific objectives to tactical level units.

Tactical Level:

The Tactical level directs operational activities toward specific objectives. Tactical level officers include Branch Directors, Division, Group, and Sector Supervisors who are in charge of grouped resources. Tactical level officers are responsible for specific geographic areas or functions, and supervising assigned personnel. A tactical level assignment comes with the authority to make decisions and assignments, within the boundaries of the overall plan and safety conditions. The accumulated achievements of tactical objectives should accomplish the strategy as outlined in the Incident Action Plan.

Task Level:

The Task Level refers to those activities normally accomplished by individual companies or specific personnel. The task level is where the work is actually done. Task level activities are routinely supervised by Company Officers. The accumulated achievements of task level activities should accomplish tactical objectives.

Command Structure - Basic Organization

Examples:

The most basic Command structure combines all three levels of the Command structure. The Company Officer on a single engine response to a dumpster fire determines the strategy and tactics, and supervises the crew doing the task.

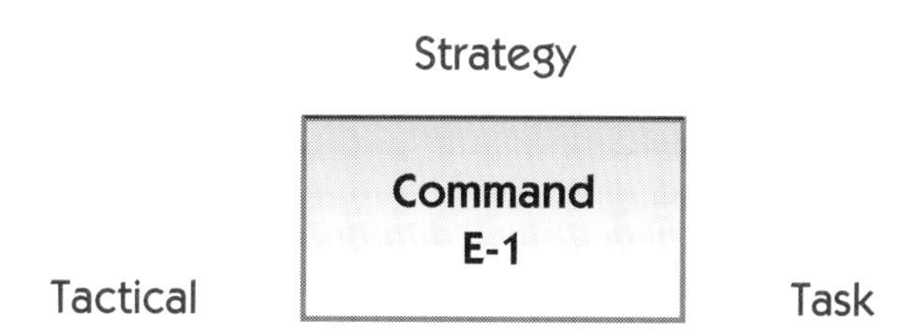

The basic structure for a "routine" incident, involving a small number of companies, requires only two levels of the Command structure. The role of Command combines the strategic and tactical levels. Companies report directly to Command and operate at the task level.

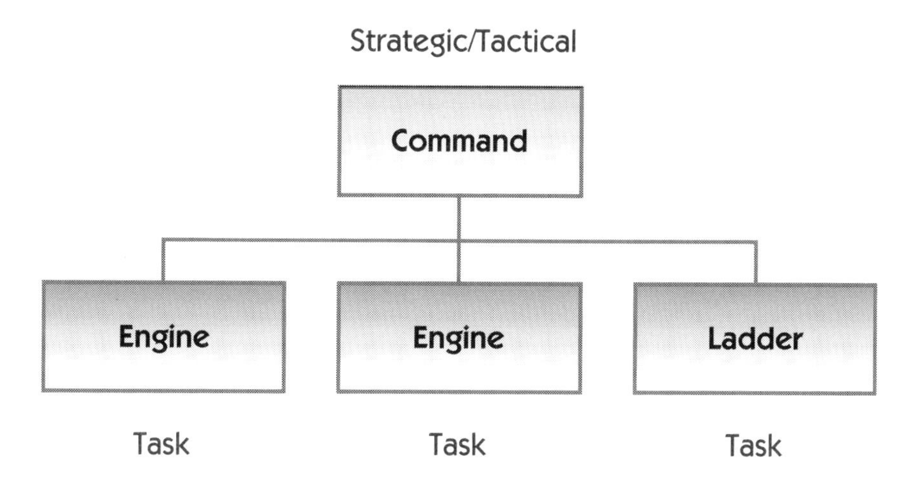

Command Structure (Division/Group, or Sector)

The terms **Divisions**, **Groups**, or **Sectors** are tactical level management units that group companies. Divisions represent geographic operations, and groups represent functional operations. The term sector is generic and can be used for both geographic and functional operations. Since all three terms are in common use throughout the country, the local jurisdiction should select the term they prefer. The following examples illustrate the use of these terms.

Tactical Level Officers: (Division/Group, or Sector)

As an incident escalates, the Incident Commander should group companies to work in Divisions/Groups, or Sectors. A Division is the organizational level having responsibility for operations within a defined geographic area. In order to effectively use the Division terminology, a department must have a designated method of dividing an incident scene.

Division Designation

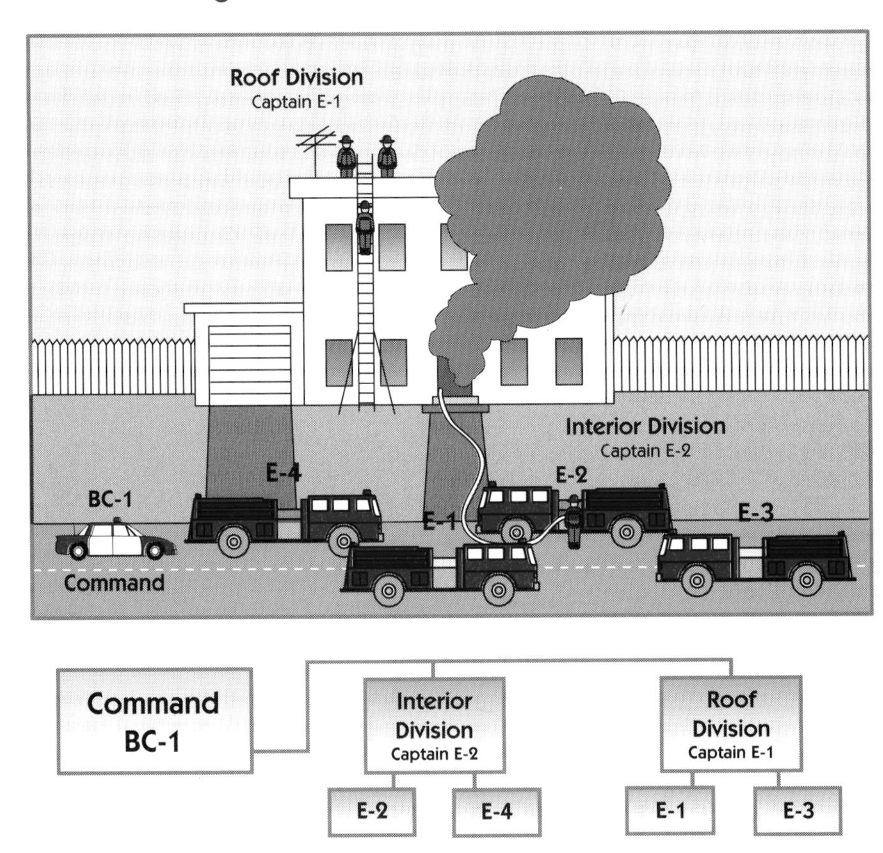

Sector Designation

Sector is either a geographic or functional assignment. Sector may take the place of either the division or group terminology or both. A fire department must determine which one or two of the three terms are to be used in their specific system. The organization structure remains the same.

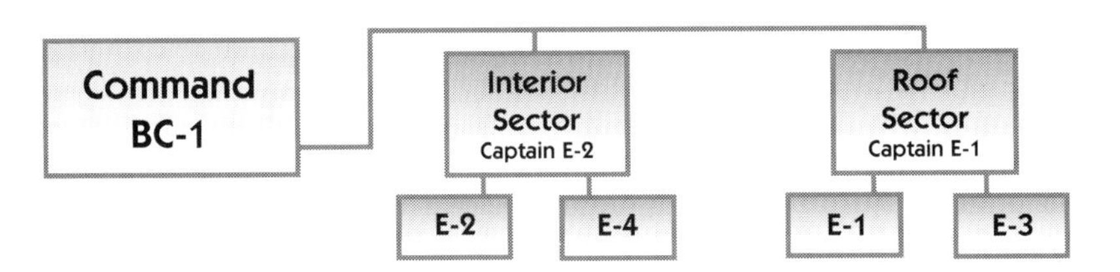

Division/Sector Designation

Tactical Assignments for a Multi-Story Incident

In multi-story occupancies, divisions will usually be indicated by floor number (Division 6 indicates 6th floor). When operating in levels below grade such as basements, the use of subdivisions is appropriate.

Division Designation

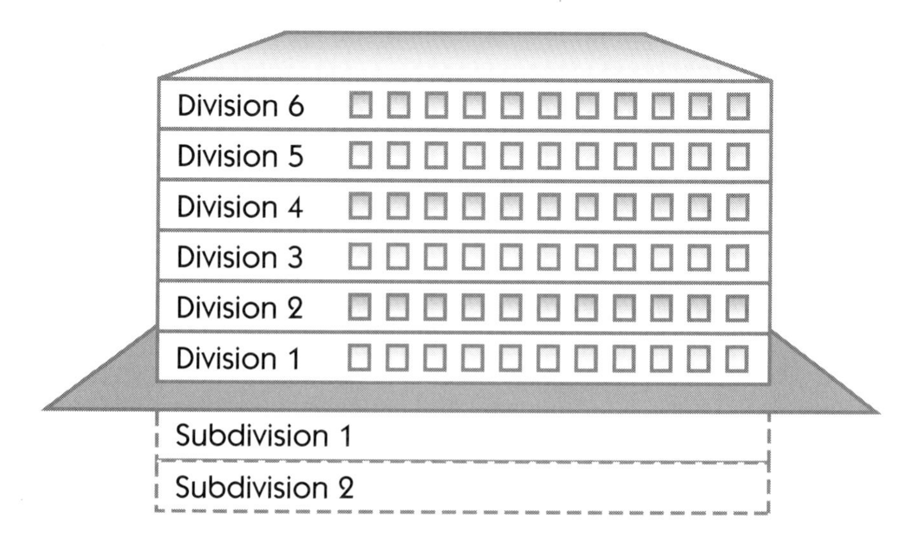

Sector Designation

Tactical Assignments for a Multi-Story Incident

In multi-story incidents, sectors will usually be indicated by floor number (Sector 6 indicates 6th floor). When operating in levels below grade such as basements, the use of subsectors is appropriate.

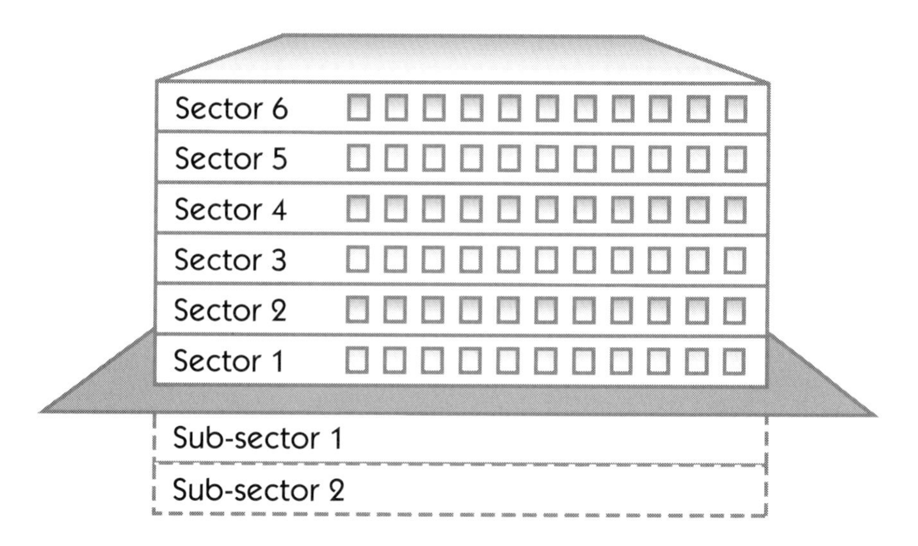

Division/Sector Designation

For situations where the incident has an odd geographical layout—not obvious North, South, East and West—the front of the building is designated "Division A" or "Sector A," and the remaining sides are given a radio designation of B, C, D in a **clockwise** manner.

Exterior designations are identified by alphabetical letter identifiers. Starting at the front of a building and progressing clockwise around the building as illustrated. Division A or Sector A will **always** indicate the **front** of the building.

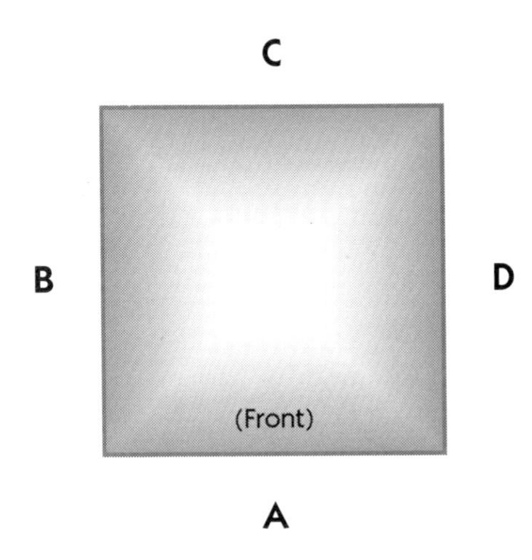

For example:

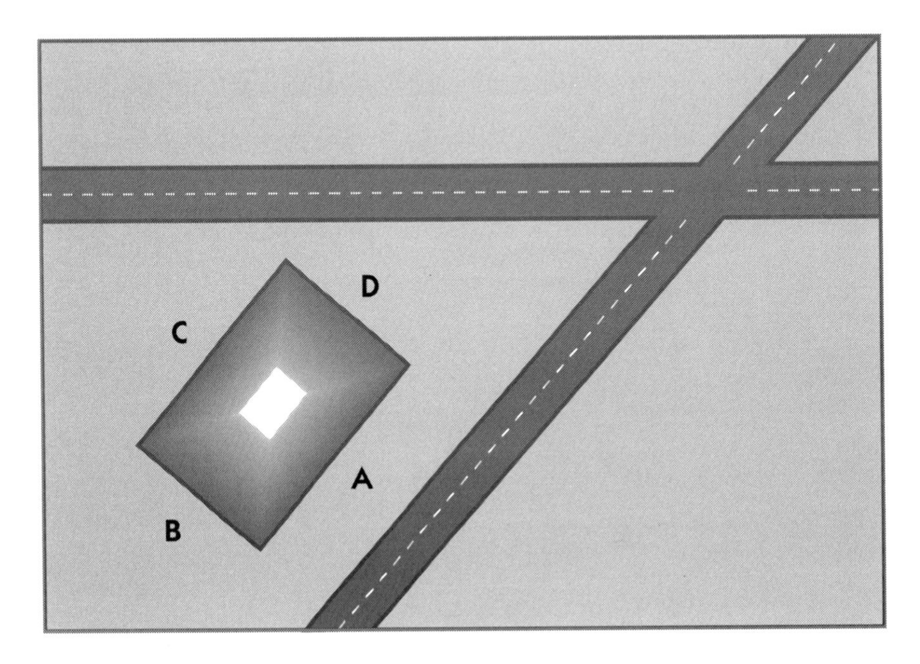

Note: For clarity of purpose during radio communications, the phonetic designations of "Alpha," "Beta," "Charlie" and "Delta" are suggested. For example "Delta sector to Command."

Division/Group/Sector Designation

A division is that organization level having responsibility for operations within a defined geographic area. The Division level is organizational between Single Resources, Task Force, or the Strike Team and the Branch.

Groups are an organizational level responsible for a specific functional assignment at an incident. Examples are Salvage Group, Search and Rescue Group, Haz Mat Group and Medical Group.

Division/Group Designation

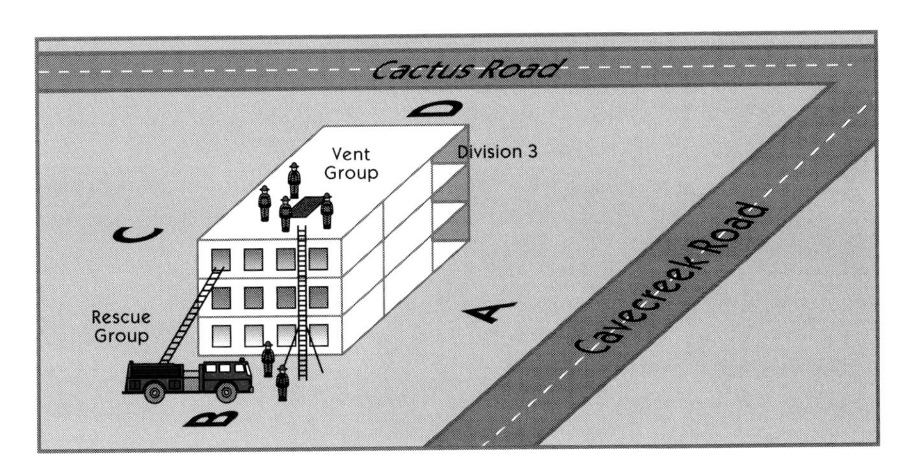

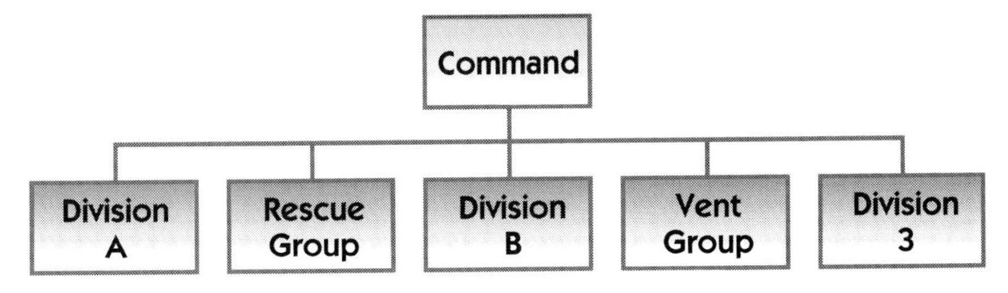

Sector Designation

Sector is either a geographic or functional assignment. Sector may take the place of either division or group terminology or both. A fire department must determine which one or two of the three terms are to be used in their specific system. The organization structure remains the same.

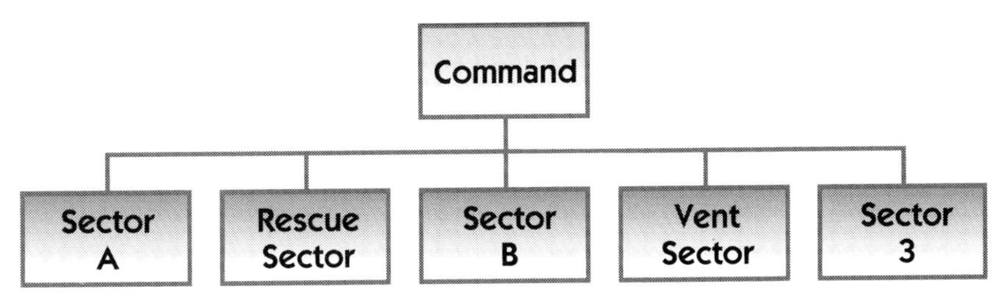

Command Structure - Division/Group or Sector; Basic Operational Approach

The use of Divisions/Groups, or Sectors in the Command organization provides a standard system to divide the incident scene into smaller subordinate management units or areas.

Complex emergency situations often exceed the capability of one officer to effectively manage the entire operation. Divisions/ Groups or Sectors reduce the span of control to more manageable smaller-sized units. Divisions/Groups or Sectors allow the Incident Commander to communicate principally with these organizational levels, rather than multiple, individual Company Officers providing an effective Command structure and incident scene organization. Generally, Division/Group or Sector responsibilities should be assigned early in the incident, **typically to the first Company assigned to a geographic area or function**. This early establishment of Division/ Group or Sector provides an effective Incident Command organization framework on which the operation can be built and expanded.

The number of Divisions/Groups or Sectors that can be effectively managed by the Incident Commander varies. Normal span of control is 3-7. In fast-moving, complex operations, a span of control of no more than 5 Divisions/Groups, or Sectors is indicated. In slower-moving less complex operations, the Incident Commander may effectively manage more Divisions/Groups, or Sectors.

Where the number of Divisions/Groups or Sectors exceeds the span-of-control that the Incident Commander can effectively manage, the incident organization can be expanded to meet incident needs by assigning an Operations Section Chief. The Operations Section is responsible for the Branches, Divisions/Groups or Sectors. Each Branch is responsible for several Divisions/Groups or Sectors and should be assigned a separate radio channel if available. **(See Branches on page 25 and Operations Section on page 33.)**

Division/Group or Sector procedures provide an array of major functions which may be selectively implemented according to the needs of a particular situation. This places responsibility for the details and execution of each particular function on a Division/Group or Sector.

When effective Divisions/Groups or Sectors have been established, the Incident Commander can concentrate on overall strategy and resource assignment, allowing the Divisions/Groups or Sectors to manage their assigned units. The Incident Commander determines strategy and assigns tactical objectives and resources to the Divisions/ Groups or Sectors. Each Division/Group or Sector Supervisor is responsible for the tactical deployment of the resources at their disposal, in order to complete the tactical objectives assigned by the Incident Commander. Divisions/Groups or Sectors are also responsible for communicating needs and progress to Command.

Divisions/Groups or Sectors reduce the overall amount of radio communications. Most routine communications within a Division/Group or Sector should be conducted in a face-to-face manner between Company Officers and their Division/Group or Sector. This process reduces unnecessary radio traffic and increases the ability to transmit critical radio communications.

The safety of firefighting personnel represents the major reason for establishing Divisions/Groups or Sectors. Each Division/Group or Sector must maintain communication with assigned companies to control both their position and function. The Division/Group or Sector must constantly monitor all hazardous situations and risks to personnel. The Division/Group or Sector must take appropriate action to ensure that companies are operating in a safe and effective manner.

The Incident Commander should begin to assign Divisions/Groups or Sectors based on the following factors:

- Situations which will eventually involve a number of companies or functions, beyond the capability of Command to directly control. Command should initially assign Division/Group or Sector responsibilities to the first companies assigned to a geographic area or function until Chief Officers are available.
- When Command can no longer effectively cope with (or manage) the number of companies currently involved in the operation.
- When companies are involved in complex operations (large interior or geographic area, hazardous materials, technical rescues, etc.)
- When companies are operating from tactical positions which Command has little or no direct control over (i.e., out of sight).
- When the situation presents special hazards and close control is required over operating companies (i.e., unstable structural conditions, hazardous materials, heavy fire load, marginal offensive situations, etc.).

When establishing a Division/Group or Sector, the Incident Commander will assign each Division/Group or Sector:

1. Tactical objectives
2. A radio designation (Roof Division/Sector, Division/Sector A)
3. The identity of resources assigned to the Division/Group or Sector.

Division/Group or Sector Guidelines

Divisions/Groups or Sectors will be regulated by the following guidelines:

- It will be the ongoing responsibility of Command to assign Divisions/Groups or Sectors as required for effective emergency operations; this assignment will relate to both geographic and functional Divisions/Groups or Sectors.
- Command shall advise each Division/Group or Sector of specific tactical objectives. The overall strategy and plan will and should be provided, (time permitting) so the Division/Group or Sector has some idea of what is going on and how their assignment fits into the overall plan.
- The number of companies assigned to a Division/Group or Sector will depend upon conditions within that Division/Group or Sector. Command will maintain an awareness of the number of companies operating within a Division/Group or Sector and the capability of that Division/Group or Sector to effectively direct operations. If a Division/Group or Sector cannot control the resources within the Division/Group or Sector, they should notify the Incident Commander so that Division/Group or Sector responsibilities can be split or other corrective action taken. In most cases 3-7 companies represent the maximum span of control for a Division/Group or Sector.
- The incident scene should be subdivided in a manner that makes sense. This should be accomplished by assigning Divisions/Sectors to geographic locations (i.e., Roof Sector, Division A, etc.) and assigning functional responsibilities to Groups/Sectors (i.e. Ventilation Group, Salvage Sector, etc.).

Division/Group or Sector Supervisors will use the Division/Group or Sector designation in radio communications (i.e., "Roof Sector to Command").

Divisions/Groups or Sectors will be commanded by Chief Officers, Company Officers, or any other Fire Department member designated by Command.

The guideline for span-of-control with Divisions/Groups or Sectors is five. This applies to Operational Division/Group or Sectors. Many of the functional responsibilities (P.I.O., Safety, etc.) are preassigned to certain individuals and are driven by standard operating procedures. These types of functional responsibilities should operate automatically and as such should not be included in the Incident Commander's span of control.

Regular Transfer of Command procedures will be followed in transferring Division/Group or Sector responsibility.

In some cases, a Division/Group or Sector Supervisor may be assigned to an area/function initially to evaluate and report conditions and advise Command of needed tasks and resources. The assigned officer will proceed to the Division/Group or Sector, evaluate and report conditions to the Incident Commander, and assume responsibility for directing resources and operations within his/her assigned area of responsibility.

The Division/Group or Sector Supervisor must be in a position to directly supervise and monitor operations. This will require the Division/Group or Sector Supervisor to be equipped with the appropriate protective clothing and equipment for their area of responsibility. Division/Group or Sector Supervisors assigned to operate within the hazard zone must be accompanied by a partner.

Division/Group or Sector Supervisors will be responsible for and in control of all assigned functions within their Division/Group or Sector. **This requires each Division/Group or Sector Supervisor to:**

A. Complete objectives assigned by Command.

B. Account for all assigned personnel.

C. Ensure that operations are conducted safely.

D. Monitor work progress.

E. Redirect activities as necessary.

F. Coordinate actions with related activities, and adjacent Divisions/Groups or Sectors.

G. Monitor welfare of assigned personnel.

H. Request additional resources as needed.

I. Provide Command with essential and frequent progress reports.

J. Re-allocate resources within the Division/Group or Sector.

The Division/Group or Sector Supervisor should be readily identifiable and maintain a visible position as much as possible.

The primary function of Company Officers working within a Division/Group or Sector is to direct the operations of their individual crews in performing assigned tasks. Company Officers will advise their Division/Group or Sector Supervisor of work progress, preferably face-to-face. All requests for additional resources or assistance within a Division/Group or Sector must be directed to the Division/Group or Sector Supervisor. Division/Group or Sector Supervisors will communicate with "Command."

Each Division/Group or Sector Supervisor will keep Command informed of conditions and progress in the Division/Group or Sector through regular progress reports. The Division/Group or Sector Supervisor must prioritize progress reports to essential information only.

Command must be advised immediately of significant changes, particularly those involving the ability or inability to complete an objective, hazardous conditions, accidents, structural collapse, etc.

When a company is assigned from Staging to an operating Division/Group or Sector, the company will be told to what Division/Group or Sector, and the name of the Supervisor they will be reporting to. The Division/Group or Sector Supervisor will be informed of which particular companies or units have been assigned by the Incident Commander. It is then the responsibility of the Division/Group or Sector Supervisor to contact the assigned company to transmit any instructions relative to the specific action requested.

Division/Group or Sector Supervisors will monitor the condition of the crews operating in their Sector. Relief crews will be requested in a manner to safeguard the safety of personnel and maintain progress toward the Division/Group or Sector objectives.

Division/Group or Sector Supervisors will ensure an orderly and thorough reassignment of crews to Responder Rehab. Crews must report to Responder Rehab intact to facilitate accountability.

3

Command Structure-Expanding the Organization

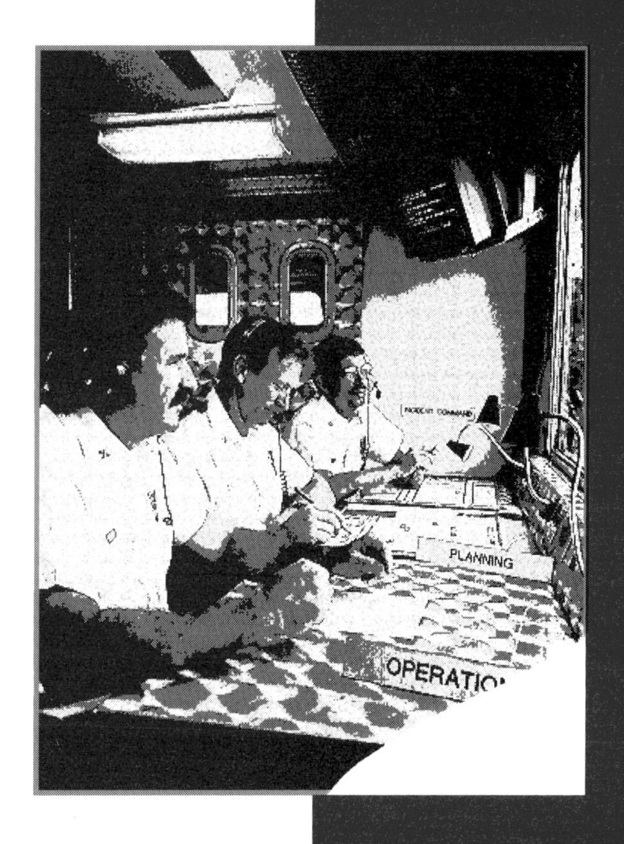

3

Command Structure–Expanding the Organization

As a small incident escalates into a major incident, the span of control may become stretched as more divisions/groups or sectors are implemented. In addition, the Incident Commander can become quickly overwhelmed and overloaded with Information management, assigning companies, filling out and updating the tactical worksheets, planning, forecasting, requesting additional resources, talking on the radio, and fulfilling all the other functions of Command. The immediate need of the Incident Commander is support. As additional ranking officers arrive on the scene, the Command organization may be expanded through implementation of branches and sections and the involvement of Officers and staff personnel to fill Command and General Staff Positions.

Section level positions can be implemented at any time, based on the needs of the incident. One of the first sections typically implemented is the Operations Section Chief.

Operations Section - Overview

The Operations Section is responsible for the direct management of all incident tactical activities, the tactical priorities, and the safety and welfare of the personnel working in the Operations Section. The Operations Section Chief uses the appropriate radio channel to communicate strategic and specific objectives to the Branches and/or Divisions/Groups or Sectors.

The Operations Section is most often implemented (staffed) as a span-of-control mechanism. When the number of Branches, Divisions/Groups or Sectors exceeds the capability of the Incident Commander to effectively manage, the Incident Commander may staff the Operations Section to reduce their span of control and thus transfer direct management of all tactical activities to the Operations Section Chief. The Incident Commander is then able to focus their attention on management of the entire incident rather than concentrating on tactical activities. **The Operations Section Chief responsibilities will be discussed in more detail beginning on page 33.**

Expanding the Organization - Branches

As previously discussed in this procedure, Divisions/Groups or Sectors identify tactical level assignments in the Command Structure. As the span-of-control begins to be excessive, the incident becomes more complex, or has two or more distinctly different operations (i.e., Fire, Medical, Evacuation, etc.), the organization can be further sub-divided into Branches.

Branches may be established on an incident to serve several purposes. However, they are not always essential to the organization of the Operations Section.

In general, branches may be established for the following reasons:

- Span of Control
- Functional
- Multi-Jurisdictional
- **When the numbers of Divisions/Groups or Sectors exceed the recommended span of control for the Operations Section Chief.** The Incident Commander or Operations Section Chief should designate a Multi-Branch structure, and allocate the Divisions/ Groups or Sectors within those Branches.

In the following example, the Operations Section Chief has one Group and four Divisions reporting to him and two additional Divisions and one Group being added. At this point, a two-Branch organization was formed, as reflected below and on the next page.

Before Multi-Branch structure

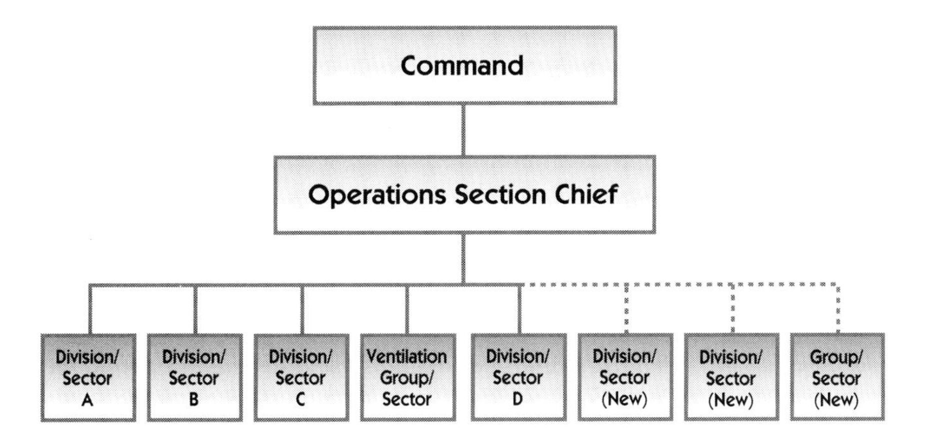

Two-Branch Organization:

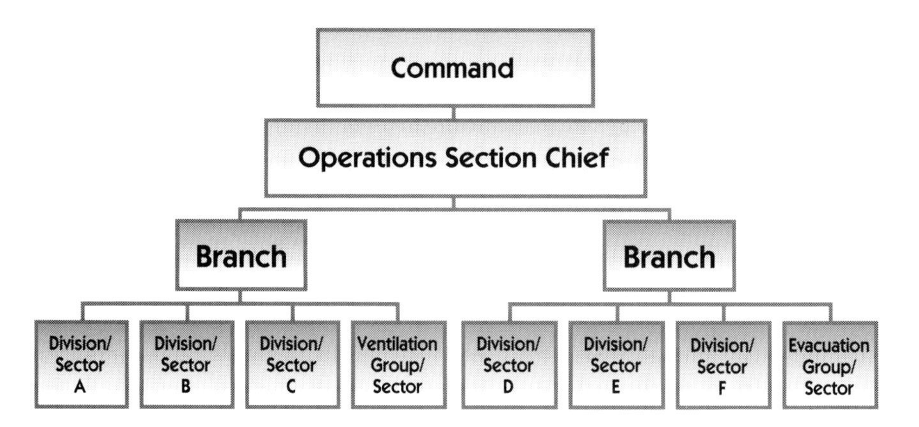

Branches should operate in their area of responsibility on separate radio channels and communicate to Operations on a different channel if possible. The radio designation of Branches should reflect the objective of the Branch, when designating functional branches, (i.e., Haz Mat Branch, Multi-Casualty Branch, etc.). Tactical Branches may be designated numerically (i.e., Branch I, Branch II, Branch III, etc.). When Operations implements Branch Directors, the Division/ Group or Sector Supervisors should be notified of their new supervisor. This Information should include:

1. What Branch the Division/Group or Sector is now assigned to.
2. The radio channel the Branch (Division/Group or Sector) is operating on.

Radio Communications should then be directed from the Division/Group or Sector Supervisor to the Branches - instead of Command orOperations. Branch Directors will receive direction from Command or Operations, which will then be relayed to Divisions/ Groups or Sectors.

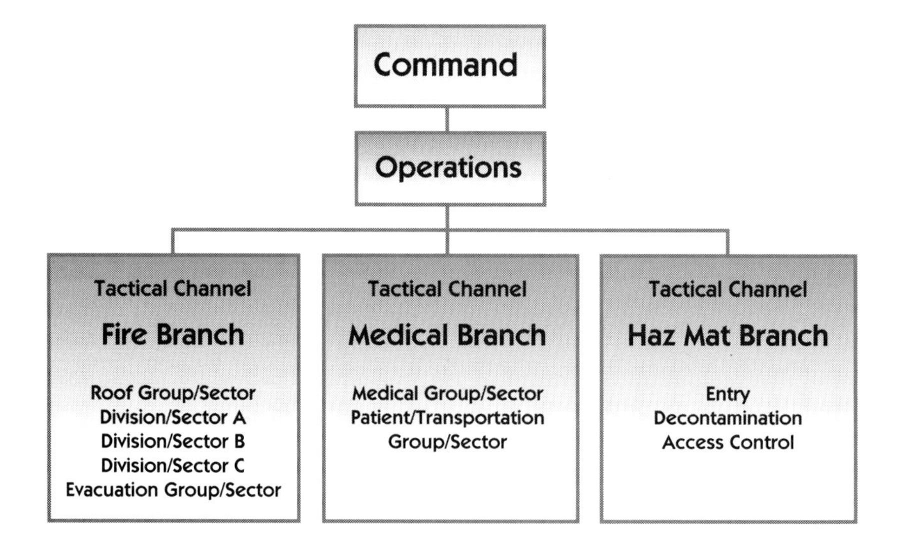

Depending on the situation, Branches may be located at the Command Post or at operational locations. When located at the Command Post, Branches can communicate on a face-to-face basis with the Operations Section Chief and/or Incident Commander. When an incident encompasses a large geographic area, it may be more effective to have Branches in tactical locations. When Branches are sent to tactical positions they should immediately implement Command and control procedures within their Branch. In these situations Operations must assign someone to monitor a "Command Channel."

Branches are not limited to Operations. Any of the Section Chiefs may recommend the implementation of Branches within their sections with approval of the Incident Commander.

Organization expands from this...

Incident Commander
- Evacuation Group/Sector
- Roof Division/Sector
- Division/Sector A
- Division/Sector B
- Division/Sector C
- Haz Mat Group/Sector
- Patient Transport. Group/Sector
- Medical Group/Sector

to this.

Incident Commander
- Operations Section Chief
  - Fire Branch
    - Evacuation Group/Sector
    - Roof Division/Sector
    - Division Sector A
    - Division Sector B
    - Division Sector C
  - Haz Mat Branch
    - Entry
    - Decon
    - Access Control
  - Multi-Casualty Branch
    - Patient Transportation Group/Sector
    - Medical Group/Sector

Functional Branch Structure

When the nature of the incident calls for a functional Branch structure, i.e., a major aircraft crash within a jurisdiction, three departments within the jurisdiction (police, fire, and health service), each has a functional Branch operating under the direction of a single Operations Section Chief. In this example, the Operations Section Chief is from the fire department with deputies from police and health services departments. Other alignments could be made depending upon the jurisdiction plan and type of emergency. Note that Incident Command in this situation could be either Single or Unified Command depending upon the jurisdiction.

Functional Branches

Command

Operations Section Chief (Fire)

Deputy Law Enforcement

Deputy Health Services

Law Branch

Fire Branch

Medical Branch

Multi-jurisdictional Incidents

When the incident is multi-jurisdictional, resources are best managed under the agencies which have normal control over those resources.

Branches should be utilized at incidents where the span of control with Divisions/Groups or Sectors is maximized, incidents involving two or more distinctly different major management components (i.e., a large fire with a major evacuation, a large fire with a large number of patients). The Incident Commander may elect to assign Branches to forward positions to manage and coordinate activities, as illustrated on the next page.

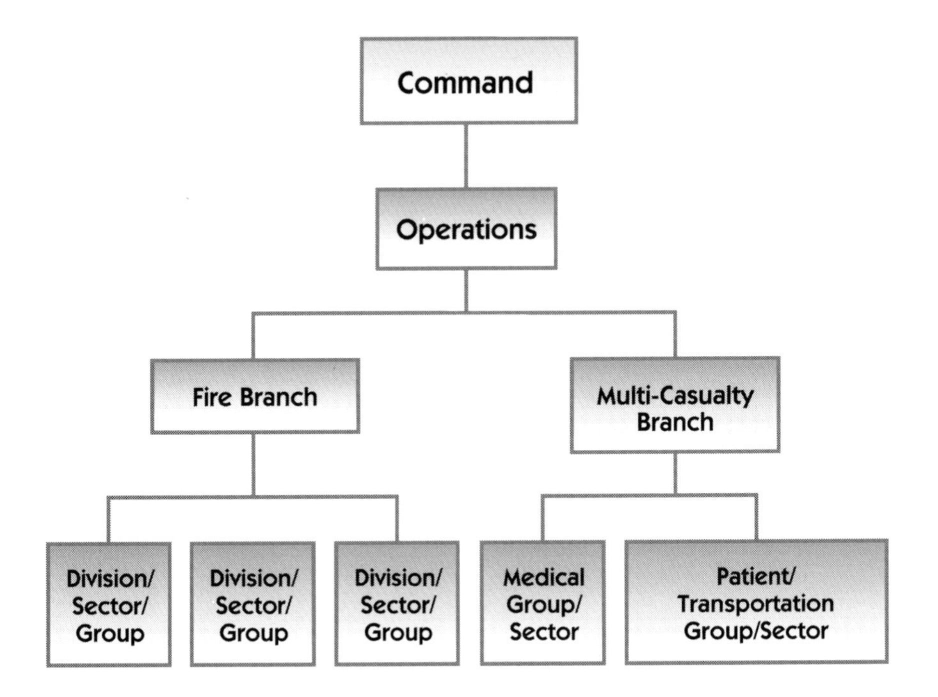

When the incident requires the use of aircraft, such as for the transportation of victims from a multi-casualty incident, highrise roof top rescue, swift water rescue, or wildland fire, the Operations Section Chief should establish the Air Operations organization. Its size, organization, and use will depend primarily upon the nature of the incident, and the availability of aircraft.

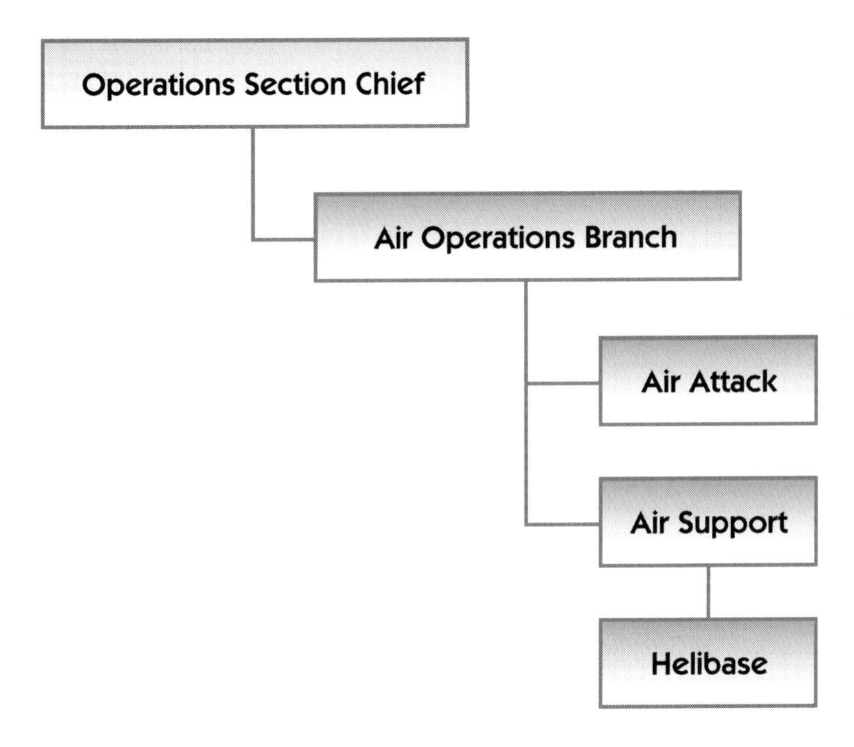

Expanding the Incident Command Organization

As the organization expands to deal with a major incident, the Incident Commander will need additional Command Post support. The Operations Section Chief (see page 33) is one of the first to be implemented.

The following organizational charts are examples of how the Incident Management System can expand to fit the size and complexity of various types of incidents.

Command Procedures - Expanding the Organization

Structure Fire

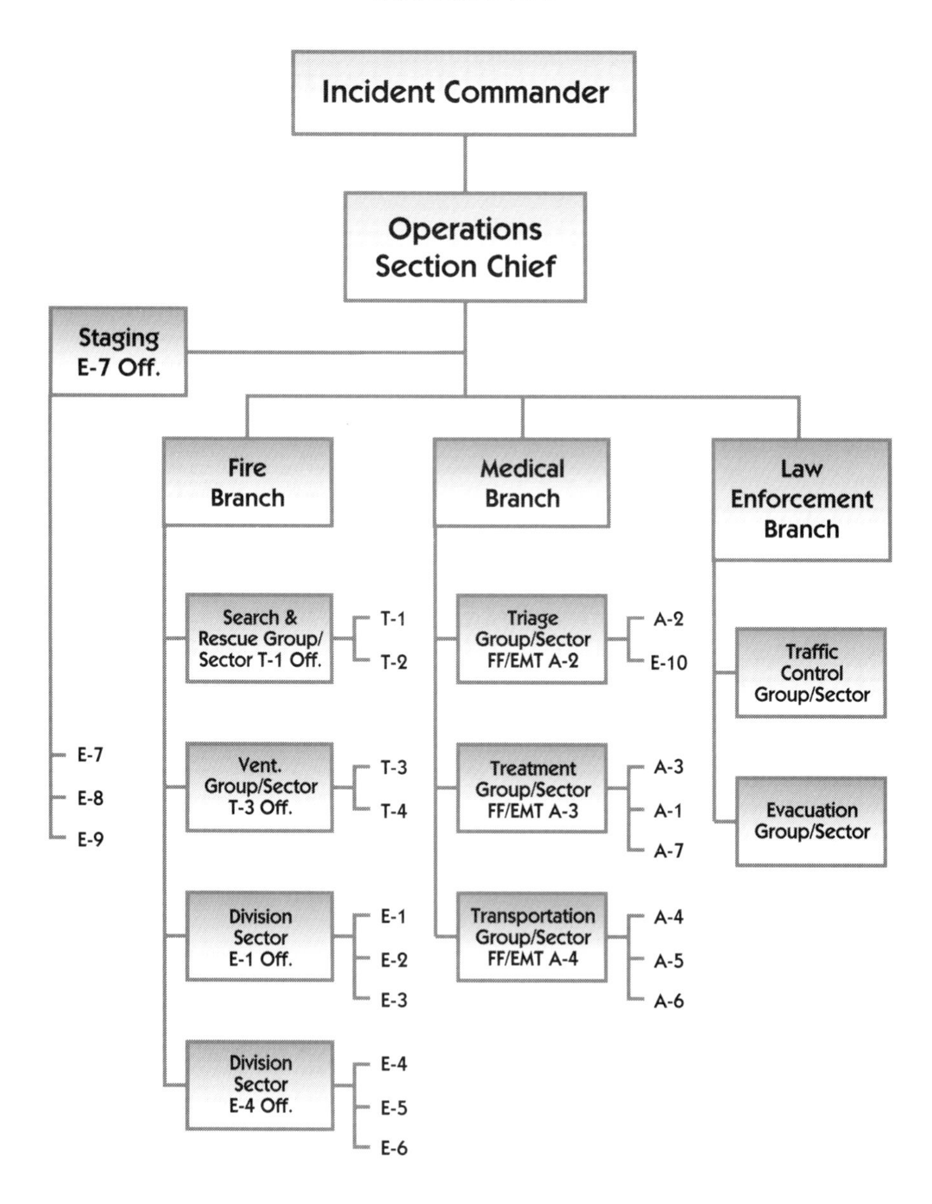

Organizational Hierarchy

The Incident Management System organizational structure develops in a modular fashion based upon the kind and size of an incident. The organization's staff builds from the top down with responsibility and performance placed initially with the Incident Commander. As the need exists, four separate Sections can be developed, each with several Units which may be established. The specific organization structure established for any given incident will be based upon the management needs of the incident. If one individual can simultaneously manage all major functional areas, no further organization is required. If one or more of the areas requires independent management, an individual is named to be responsible for that area.

For ease of reference and understanding, personnel assigned to manage at each level of the organization will carry a distinctive organizational title.

- **Command**
- **Officer**
- **Section Chiefs**
- **Director**
- **Supervisor**
- **Unit Leader**
- **Managers**
- **Single Resources**

Command: Refers to the Incident Commander.

Officer: Title that refers to a member of the Command Staff (Information Officer, Safety Officer, Liaison Officer).

Section Chiefs: Title that refers to a member of the General Staff (Planning Section Chief, Operations Section Chief, Finance/Administration Section Chief, Logistics Section Chief).

Directors: Title that refers to the positions of Branch Director, which is in the Operations Section, or Logistics Section between the Divisions/Groups or Sectors, and the Operations Section Chiefs (Branch Directors, Air Operations Branch Director, Service Branch Director).

Supervisors: Title that refers to the positions of Division/Group or Sector Supervisor, which is in the Operations Section and lies between the Branch Director and Strike Team/Task Force Leader.

Unit Leader: Title that refers to a position with supervision and management responsibility of either a group of resources or a unit, such as Ground Support, Medical, Supply, etc.

Managers: Title that refers to the lowest level of supervision within the Logistics Section: Equipment Manager, Base Manager, Camp Manager. The only exception to this is the Staging Area Manager who reports directly to the Operations Section Chief.

Single Resources: Engine company, truck company, with a company officer and crew.

Command Structure - Expanding the Organization "Sections"

As previously noted, as a small incident escalates into a major incident, additional organizational support will be required. The Incident Commander can become quickly overwhelmed and overloaded with Information management, assigning companies, filling out and updating the tactical worksheets, planning, forecasting, requesting additional resources, talking on the radio, and fulfilling all the other functions of Command. The immediate need of the Incident Commander is support. As additional ranking officers arrive on the scene, the Command organization may be expanded through the involvement of Officers and staff personnel to fill Command and General Staff Positions.

Section and Unit level positions within the Incident Management System will be activated only when the corresponding functions are required by the incident.

Until such time as a Section or Unit is activated, all functions associated with that Section or Unit will be the responsibility of the Incident Commander or the appropriate Section Chief. It is recommended that two or more units **not** be combined into a single unit. However, an individual may be assigned responsibility for managing more than one unit. This method of organization allows for easy expansion and demobilization of the system.

The Command structure defines the lines of authority, but it is not intended that the transfer of information within the Incident Management System be restricted to the chain of Command. An individual will receive **orders** from a superior, but may give information to any position in a different part of the organization within the guidelines specified in the operational procedures for each position.

The majority of positions within the Incident Management System will not be activated until the initial response is determined to be insufficient to handle the situation. When this occurs, qualified personnel are requested through normal dispatching procedures to fill the positions determined to be required for the type of incident in progress. If it is later determined that a specific position is not needed, the request can be canceled. Some agencies have elected to use a modular form of dispatching; e.g., entire units.

The transition from the initial response to a major incident organization will be evolutionary and positions will be filled as the corresponding tasks are required.

During the initial phases of the incident the Incident Commander normally carries out these four section functions.

1. **Operations**
2. **Planning**
3. **Logistics**
4. **Finance/Administration**

These comprise the General Staff within a fully expanded incident organizational structure.

Expanding the Organization– Sections

Section level positions can be implemented at any time, based on the needs of the incident. One of the first sections typically implemented is the Operations Section Chief.

Operations Section

The Operations Section is responsible for the direct management of all incident tactical activities, the tactical priorities, and the safety and welfare of the personnel working in the Operations Section. The Operations Section Chief uses the appropriate radio channel to communicate strategic and specific objectives to the Branches and/ or Divisions/Groups or Sectors.

The Operations Section is most often implemented (staffed) as a span-of-control mechanism. When the number of Branches, Divisions/Groups or Sectors exceeds the capability of the Incident Commander to effectively manage, the Incident Commander may staff the Operations Section to reduce their span-of-control and thus transfer direct management of all tactical activities to the Operations Section Chief. The Incident Commander is then able to focus their attention on management of the entire incident rather than concentrating on tactical activities.

Operations Section Chief

The Incident Operations Section Chief is responsible for the direct management of all incident tactical activities and should have direct involvement in the preparation of the action plan for the period of responsibility.

Roles and Responsibilities:

- Manage incident tactical activities.
- Coordinate activities with the Incident Commander.
- Implement the Incident Action Plan.
- Assign resources to tactical level areas based on tactical objectives and priorities.
- Build an effective organizational structure through the use of Branches and Divisions/Groups or Sectors.
- Provide tactical objectives for Divisions/Groups or Sectors.
- Control Staging and Air Operations.
- Provide for life safety.
- Determine needs and request additional resources.
- Consult with and inform other Sections and the Incident Command Staff as needed.

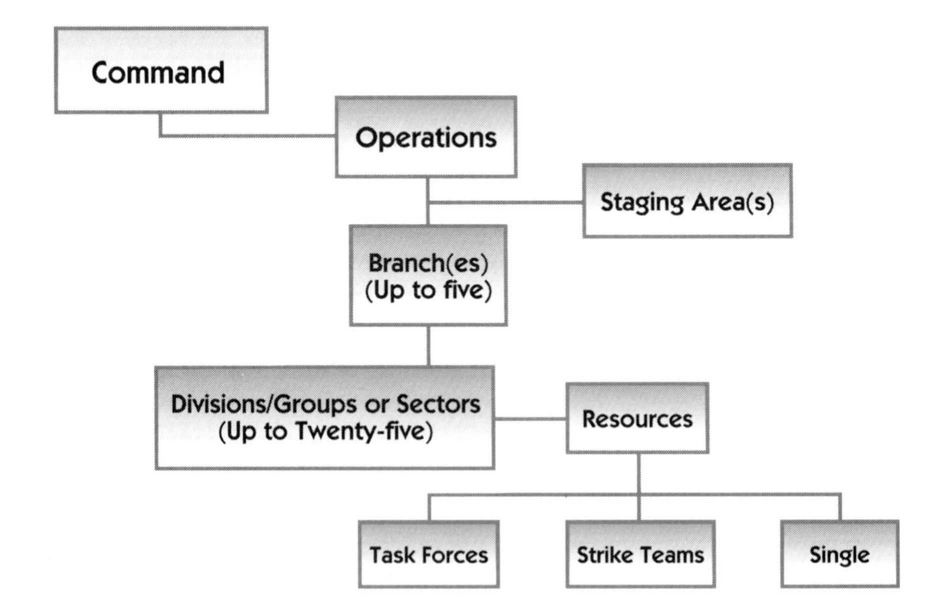

Staging Areas

Staging Areas are locations designated within the incident area which are used to temporarily locate resources which are available for assignment.

The incident scene can quickly become congested with emergency equipment if this equipment is not managed effectively. For major or complex operations, the Incident Commander should establish a central Staging Area early and place an officer in change of Staging. A radio designation of "Staging" should be utilized.

In this expanded organizational structure Staging reports to the Operations Section Chief. The Operations Section Chief may establish, move and discontinue the use of Staging Areas. All resources within the designated Staging Areas are under the direct control of the Operations Section Chief and should be immediately available. Staging will request logistical support (e.g., food, fuel, sanitation) from the Logistics Section.

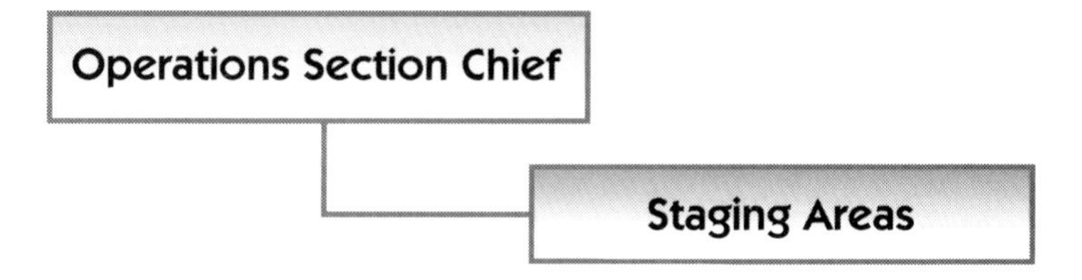

Planning Section

The Planning Section is responsible for gathering, assimilating, analyzing, and processing information needed for effective decision making. Information management is a full-time task at large and complex incidents. The Planning Section serves as the Incident Commander's "clearing house" for information. This allows the Incident Commander's staff to provide information instead of having to deal with dozens of information sources. Critical information should be immediately forwarded to Command (or whoever needs it). Information should also be used to make long range plans. The Planning Section Chief's goal is to plan ahead of current events and to identify the need for resources before they are needed.

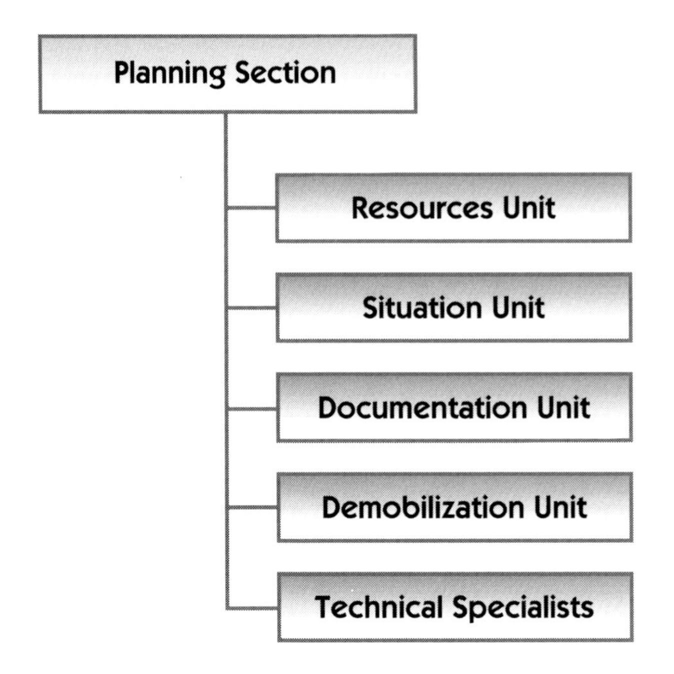

Roles and Responsibilities:

- Evaluate current strategy and plan with the Incident Commander.
- Maintain resource status and personnel accountability.
- Refine and recommend any needed changes to plan with Operations input.
- Evaluate incident organization and span-of-control.
- Forecast possible outcome(s).
- Evaluate future resource requirements.
- Utilize technical assistance as needed.
- Evaluate tactical priorities, specific critical factors, and safety.
- Gather, update, improve, and manage situation status with a standard systematic approach.
- Coordinate with any needed outside agencies for planning needs.
- Plan for incident demobilization.
- Maintain incident records.

Logistics Section

The Logistics Section is the support mechanism for the organization. Logistics provides services and support systems to all the organizational components involved in the incident including facilities, transportation, supplies, equipment maintenance, fueling, feeding, communications, and medical services, including Responder Rehab.

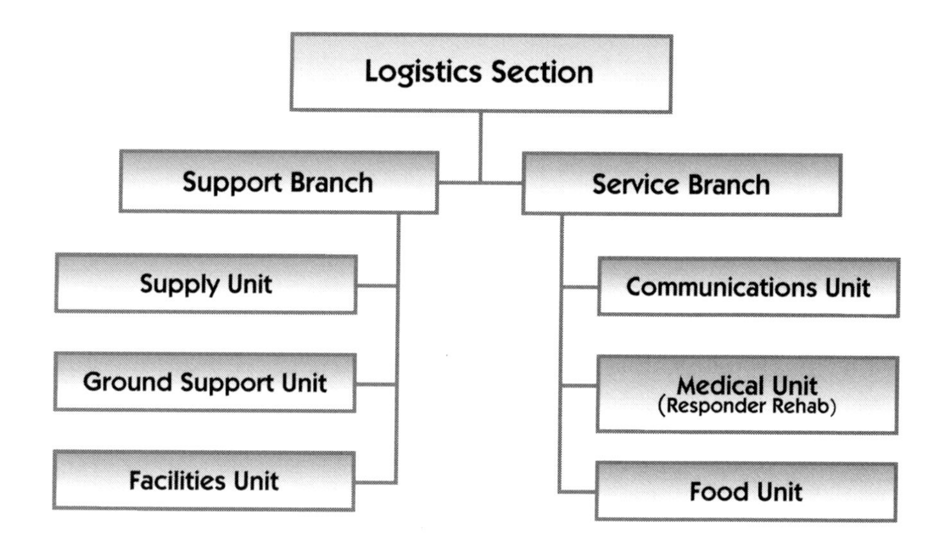

Roles and Responsibilities:

- Provide for medical aid for incident personnel and manage Responder Rehab.
- Coordinate immediate critical incident stress debriefing function.
- Provide and manage any needed supplies or equipment.
- Forecast and obtain future resource needs (coordinate with the Planning Section).
- Provide for communications plan and any needed communications equipment.
- Provide fuel and needed repairs for equipment.
- Obtain specialized equipment or expertise per Command.
- Provide food and associated supplies.
- Secure any needed fixed or portable facilities.
- Provide any other logistical needs as requested by Command.
- Supervise assigned personnel.

Finance/Administration Section

The Finance/Administration Section is established on incidents when the agency(ies) who are involved have a specific need for financial services. Not all agencies will require the establishment of a separate Finance/Administration Section. In some cases where only one specific function is required, e.g., cost analysis, that position could be established as a Technical Specialist in the Planning Section.

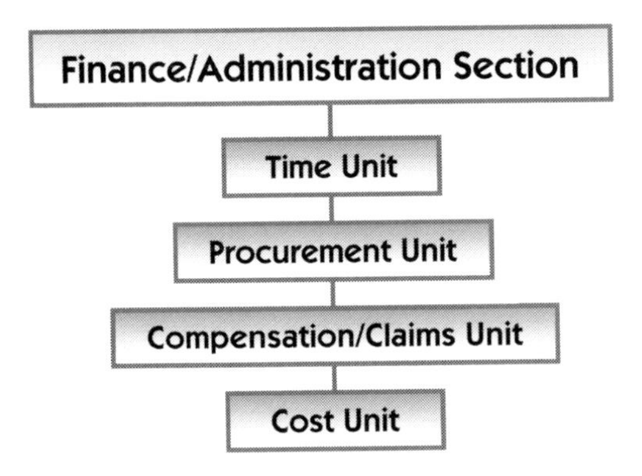

Roles and Responsibilities:

- Procuring of services and/or supplies from sources within and outside the Fire Department or City as requested by Command (coordinates with Logistics).
- Documenting all financial costs of the incident.
- Documenting for possible cost recovery for services and/ or supplies.
- Analyzing and managing legal risk for incidents (i.e., hazardous materials clean up).
- Documenting for compensation and claims for injury.

The Finance/Administration Section is responsible for obtaining any and all needed incident documentation for potential cost recovery efforts.

The Incident Commander

Role and Responsibilities **after** activation of an Operations Section Chief

Once the Operations Section is in place and functioning, the Incident Commander's focus should be on the strategic issues, overall strategic planning and other components of the incident. This focus is to look at the "big picture" and the impact of the incident from a broad perspective. The Incident Commander should provide direction, advice, and guidance to the Command and General Staff in directing the tactical aspects of the incident.

Incident Command Staff:

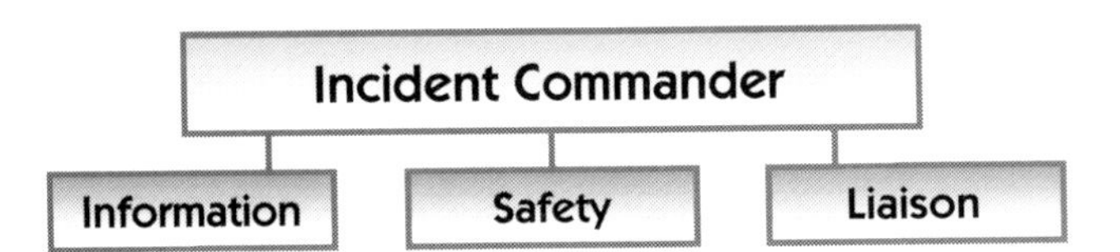

Roles and Responsibilities:

- Review and evaluate the plan, and initiate any needed changes.
- Provide on-going review of the overall incident (The Big Picture).
- Select priorities.
- Provide direction to the Command and General Staff Officer.
- Review the organizational structure, initiate change or expansion to meet incident needs.
- Stage Command and General Staff functions as necessary.
- Establish liaison with other internal agencies and officials, outside agencies, property owners and/or tenants.

Command Staff

Command staff positions are established to assume responsibility for key activities which are not a part of the line organization. Three specific staff positions are identified:

- **Information Officer**
- **Safety Officer**
- **Liaison Officer**

Additional positions might be required, depending upon the nature and location of the incident, or requirements established by Incident Command.

Information Officer

The Information Officer's function is to develop accurate and complete information regarding incident cause, size, current situation, resources committed, and other matters of general interest. The Information Officer will normally be the point of contact for the media and other governmental agencies which desire information directly from the incident. In either a single or unified Command structure, only one Information Officer would be designated. Assistants may be assigned from other agencies or departments involved.

Safety Officer

The Safety Officer's function at the incident is to assess hazardous and unsafe situations and develop measures for assuring personnel safety. The Safety Officer has emergency authority to stop and/or prevent unsafe acts. In a Unified Command structure, a single Safety Officer would be designated. Assistants may be required and may be assigned from other agencies or departments making up the Unified Command including the need for Responder Rehabilition assessment.

Liaison Officer

The Liaison Officer's function is to be a point of contact for representatives from other agencies. In a Single Command structure, the representatives from assisting agencies would coordinate through the Liaison Officer. Under a Unified Command structure, representatives from agencies not involved in the Unified Command would coordinate through the Liaison Officer. Agency representatives assigned to an incident should have authority to speak on all matters for their agency.

Expanded Organization

Incident Management - **Major Incident**

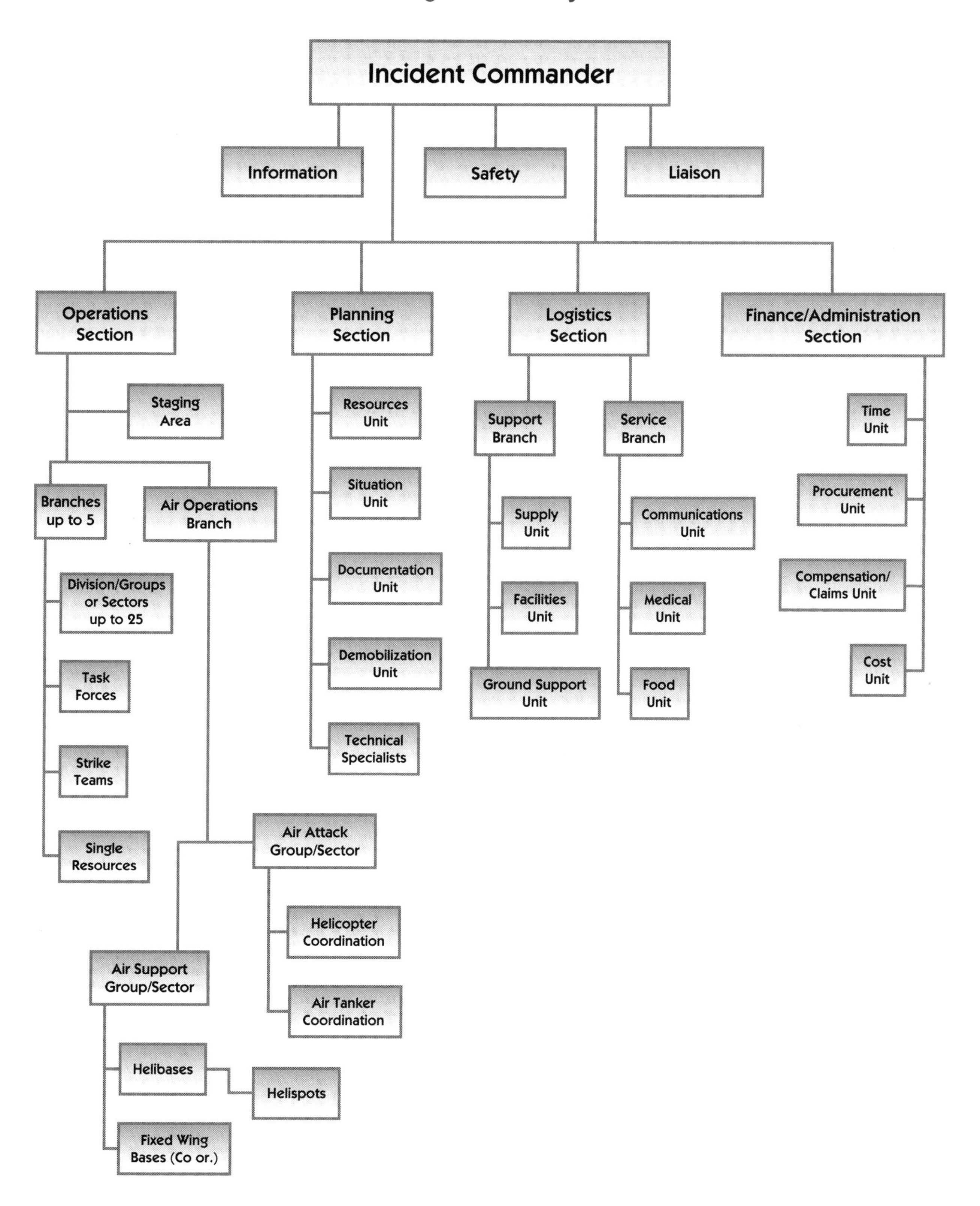

4

Unified Command

4
Unified Command

Command – Single and Unified

Command is responsible for overall management of the incident. Command also includes certain staff functions. The Command function within the Incident Management System may be conducted in two general ways.

- Single Command
- Unified Command

Single Command – Incident Commander

Within a jurisdiction in which an incident occurs, and when there is no overlap of jurisdictional boundaries involved, a single Incident Commander will be designated by the jurisdictional agency to have overall management responsibility for the incident.

The Incident Commander will prepare incident objectives which in turn will be the foundation upon which subsequent action planning will be based. The Incident Commander will approve the final action plan, and approve all requests for ordering and releasing of primary resources. The Incident Commander may have a deputy. The deputy should have the same qualifications as the Incident Commander, and may work directly with the Incident Commander, be a relief, or perform certain specific assigned tasks.

In an incident within a single jurisdiction, where the nature of the incident is primarily a responsibility of one agency, e.g., fire, the deputy may be from the same agency. In a multi-jurisdictional incident, or one which threatens to be multi-jurisdictional, the deputy role may be filled by an individual designated by the adjacent agency. More than one deputy could be involved. Another way of organizing to meet multi-jurisdictional situations is described under Unified Command.

This figure depicts an incident with Single Incident Command authority.

Single Incident Command Structure

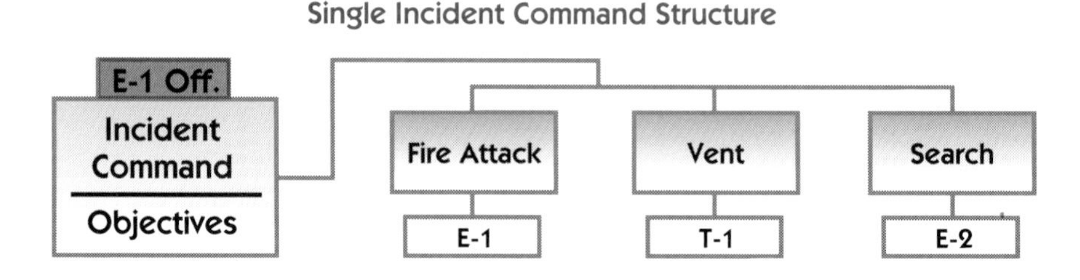

Unified Command

A Unified Command structure is called for under the following conditions:

- The incident is totally contained within a single jurisdiction, but more than one department or agency shares management responsibility due to the nature of the incident or the kinds of resources required; i.e., a passenger airliner crash within a national forest. Fire, medical, and law enforcement all have immediate but diverse objectives. An example of this kind of Unified Command structure is depicted below.

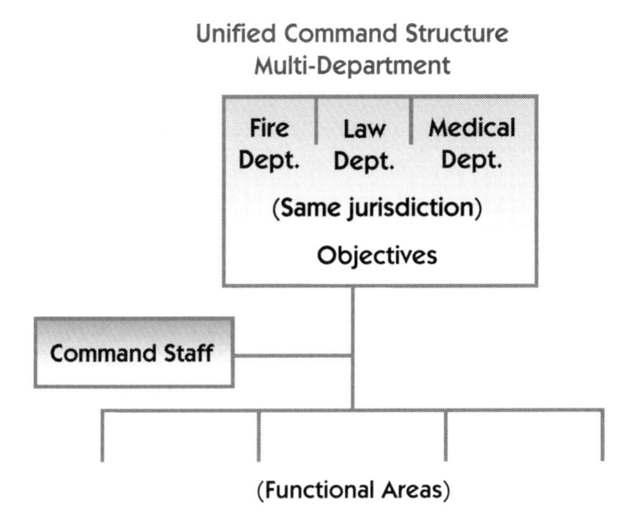

- The incident is multi-jurisdictional in nature; i.e., a major flood. An example of this Unified Command structure is shown below.

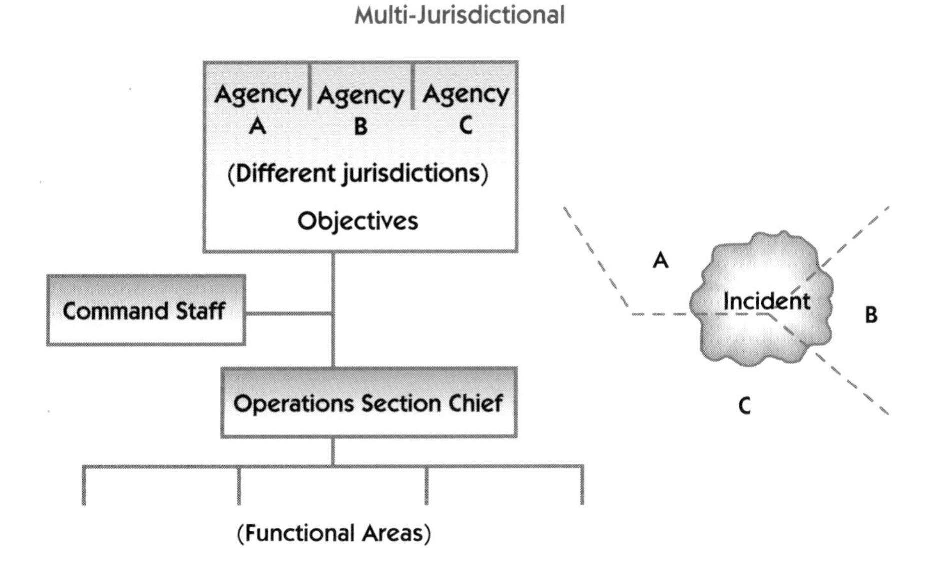

Single/Unified Command Differences

The primary differences between the Single and Unified Command structures are:

- In a Single Command structure, a single Incident Commander is solely responsible, within the confines of their authority, to establish objectives and overall management strategy associated with the incident. The Incident Commander is directly responsible for follow-through, to ensure that all functional area actions are directed toward accomplishment of the strategy. The implementation of planning required to effect operational control will be the responsibility of a single individual (Operations Section Chief) who will report directly to the Incident Commander.
- In a Unified Command structure, the individuals designated by their jurisdictions, or by departments within a single jurisdiction, must jointly determine objectives, strategy and priorities. As in a Single Command structure, the Operations Section Chief will have responsibility for implementation of the plan. The determination of which agency or department the Operations Section Chief represents must be made by mutual agreement of the Unified Command. It may be done on the basis of greatest jurisdictional involvement, number of resources involved, by existing statutory authority, or by mutual knowledge of the individual's qualifications.

Appendices

Glossary of Terms

Agency Representative – Individual assigned to an incident from an assisting or cooperating agency who has been delegated full authority to make decisions on all matters affecting that agency's participation at the incident. Agency Representatives report to the Incident Liaison Officer.

Allocated Resources – Resources dispatched to an incident that have not yet checked in with the Incident Commander.

Ambulance – A ground vehicle providing patient transport capability, specified equipment capability, and personnel (basic life support ambulance or advanced life support ambulance, etc.).

Assigned Resources – Resources checked in and assigned work tasks on an incident.

Assisting Agency – An agency directly contributing suppression, rescue, support, or service resources to another agency.

Available Resources – Resources assigned to an incident and available for an assignment.

Base – That location at which the primary logistics functions are coordinated and administered. (Incident name or other designator will be added to the term "Base.") The Incident Command Post may be co-located with the base. There is only one base per incident.

Branch – That organizational level having functional/geographic responsibility for major segments of incident operations. The Branch level is organizationally between Section and Division/Sector/Group.

Brush Patrol – A light, mobile vehicle, having limited pumping and water capacity for off-road operations.

Chief – IMS title for individuals responsible for command of the functional Sections: Operations, Planning, Logistics, and Finance/ Administration.

Clear Text – The use of plain English in radio communications transmissions. No Ten Codes or agency specific codes are used when using Clear Text.

Command Post (CP) – That location at which primary Command functions are executed; usually co-located with the Incident Base.

Command Staff – The Command Staff consists of the Information Officer, Safety Officer, and Liaison Officer, who report directly to the Incident Commander.

Command – The act of directing, ordering, and/or controlling resources by virtue of explicit legal, agency, or delegated authority.

Communications Unit – Functional Unit within the Service Branch of the Logistics Section. This unit is responsible for the incident communications plan, the installation and repair of communications equipment, and operation of the Incident Communications Center. Also may refer to a vehicle (trailer or mobile van) used to provide the major part of an Incident Communications Center.

Company Officer – The individual responsible for command of a Company. This designation is not specific to any particular fire department rank (may be a Firefighter, Lieutenant, Captain, or Chief Officer, if responsible for command of a single Company).

Company – A ground vehicle providing specified equipment capability and personnel (Engine Company, Truck Company, Rescue Company, etc.)

Compensation/Claims Unit – Functional Unit within the Finance/Administrative Section. Responsible for financial concerns resulting from injuries or fatalities at an incident.

Cooperating Agency – An agency supplying assistance other than direct suppression, rescue, support, or service functions to the incident control effort (Red Cross, law enforcement agency, telephone company, etc.).

Cost Unit – Functional Unit within the Finance/Administration Section. Responsible for tracking costs, analyzing cost data, making cost estimates, and recommending cost-saving measures.

Crew – A specific number of personnel assembled for an assignment such as search, ventilation, or hose line deployment and operations. The number of personnel in a crew should not exceed recommended span-of-control guides (3-7). A Crew operates under the direct supervision of a Crew Leader.

Demobilization Unit – Functional Unit within the Planning Section. Responsible for assuring orderly, safe, efficient demobilization of resources committed to the incident.

Director – IMS title for individuals responsible for command of a Branch.

Dispatch Center – A facility from which resources are directly assigned to an incident.

Division – That organization level having responsibility for operations within a defined geographic area. The Division level is organizational between Single Resources, Task Force, or the Strike Team and the Branch.

Documentation Unit – Functional Unit within the Planning Section. Responsible for recording/protecting all documents relevant to incident.

Engine Company – A ground vehicle providing specified levels of pumping, water, hose capacity and personnel.

Facilities Unit – Functional Unit within the Support Branch of the Logistics Section. Provides fixed facilities for incident. These facilities may include the Incident Base, feeding areas, sleeping areas, sanitary facilities, and a formal Command Post.

Finance/Administration Unit – Responsible for all costs and financial actions of the incident. Includes the Time Unit, Procurement Unit, Compensation/Claims Unit, and the Cost Unit.

Food Unit – Functional Unit within the Service Branch of the Logistics Section. Responsible for providing meals for personnel involved with incident.

General Staff – The group of incident management personnel comprised of the Operations Section Chief, Planning Section Chief, Logistics Section Chief, and Finance/Administration Section Chief.

Ground Support Unit – Functional Unit within the Support Branch of the Logistics Section. Responsible for fueling/maintaining/repairing vehicles and the transportation of personnel and supplies.

Group – That organizational level having responsibility for a specified functional assignment at an incident (ventilation, salvage, water supply, etc.)

Incident Action Plan – The strategic goals, tactical objectives, and support requirements for the incident. All incidents require an action plan. For simple incidents the action plan is not usually in written form. Large or complex incidents will require that the action plan be documented in writing.

Incident Command System (ICS) – An Incident Management System with a common organizational structure with responsibility for the management of assigned resources to effectively accomplish stated objectives pertaining to an incident.

Incident Commander (IC) – The individual responsible for the management of all incident operations.

Information Officer – Responsible for interface with the media or other appropriate agencies requiring information direct from the incident scene. Member of the Command Staff.

Initial Attack – Resources initially committed to an incident.

Ladder Company – See Truck Company.

Leader – The individual responsible for command of a Task Force, Strike Team, or Functional Unit.

Liaison Officer – The point of contact for assisting or coordinating agencies. Member of the Command Staff.

Logistics Section – Responsible for providing facilities, services, and materials for the incident. Includes the Communications Unit, Medical Unit, and Food Unit, within the Service Branch and the Supply Unit, Facilities Unit, and Ground Support Unit, within the Support Branch.

Medical Unit – Functional Unit within the Service Branch of the Logistics Section. Responsible for providing emergency medical treatment of emergency personnel. This Unit does not provide treatment for civilians.

Officer – The Command Staff positions of Safety, Liaison, and Information.

Operational Period – The period of time scheduled for execution of a given set of operation actions as specified in the Incident Action Plan.

Operations Section – Responsible for all tactical operations at the incident. Includes up to 5 Branches, 25 Divisions/Groups/Sectors, and 125 Single Resources, Task Forces, or Strike Teams.

Out-of-Service Resources – Resources assigned to an incident but unable to respond for mechanical, rest, or personnel reasons.

Planning Meeting – A meeting, held as needed throughout the duration of an incident, to select specific strategies and tactics for incident control operations and for service and support planning.

Planning Section – Responsible for the collection, evaluation, dissemination, and use of information about the development of the incident and the status of resources. Includes the Situation Status, Resource Status, Documentation, and Demobilization Units as well as Technical Specialists.

Procurement Unit – A functional Unit within the Finance/Administration Section. Responsible for financial matters involving vendors.

Reporting Locations – Any one of six facilities/locations where incident-assigned resources may check in. The locations are: Incident Command Post - Resources Unit (RESTAT), Base, Camp, Staging Area, Helibase, or Division Supervisor for direct line assignments. (Check in at one location only.)

Rescue Company – A ground vehicle providing specified rescue equipment, capability, and personnel.

Resource Status Unit (RESTAT) – Functional Unit within the Planning Section. Responsible for recording the status, and accounting of resources committed to incident and evaluation of: resources currently committed to incident, the impact that additional responding resources will have on incident, and anticipated resource needs.

Resources – All personnel and major items of equipment available, or potentially available, for assignment to incident tasks on which status is maintained.

Responder Rehabilitation (Rehab) – That function and location which shall include medical evaluation and treatment, food and fluid replenishment, and relief from extreme climatic conditions for emergency responders, according to the circumstances of the incident.

Safety Officer – Responsible for monitoring and assessing safety hazards, unsafe situations, and developing measures for ensuring personnel safety. Member of the Command Staff.

Section – That organization level having functional responsibility for primary segments of incident operations, such as: Operations, Planning, Logistics, Finance/Administration. The Section level is organizationally between Branch and Incident Commander.

Section Chief – Title that refers to a member of the General Staff (Planning Section Chief, Operations Section Chief, Finance/Administration Section Chief, Logistics Section Chief).

Sector – Is either a geographic or functional assignment. Sector may take the place of either the Division or Group or both.

Service Branch – A Branch within the Logistics Section. Responsible for service activities at incident. Components include the Communications Unit, Medical Unit, and Food Unit.

Single Resource – An individual Company or Crew.

Situation Status Unit (SITSTAT) – Functional Unit within the Planning Section. Responsible for analysis of situation as it progresses. Reports to Planning Section Chief.

Staging Area – That location where incident personnel and equipment are assigned on an immediately available status.

Strategic Goals – The overall plan that will be used to control the incident. Strategic goals are broad in nature and are achieved by the completion of tactical objectives.

Strike Team – Five (5) of the same kind and type of resources, with common communications and a leader.

Supervisor – Individuals responsible for Command of a Division/Group/Sector.

Supply Unit – Functional Unit within the Support Branch of the Logistics Section. Responsible for ordering equipment/supplies required for incident operations.

Support Branch – A Branch within the Logistics Section. Responsible for providing the personnel, equipment, and supplies to support incident operations. Components include the Supply Unit, Facilities Unit, and Ground Support Unit.

Tactical Objectives – The specific operations that must be accomplished to achieve strategic goals. Tactical objectives must be both specific and measurable. Tactical level officers are Division/Group/Sector.

Task Force – A group of any type and kind of resources, with common communications and a leader, temporarily assembled for a specific mission (not to exceed five resources).

Technical Specialists – Personnel with special skills who are activated only when needed. Technical Specialists may be needed in the areas of fire behavior, water resources, environmental concerns, resource use, and training. Technical Specialists report initially to the Planning Section but may be assigned anywhere within the IMS organizational structure as needed.

Time Unit – A functional Unit within the Finance/Administration Section. Responsible for record keeping of time for personnel working at incident.

Truck Company – A ground vehicle providing an aerial ladder or other aerial device and specified portable ladders and equipment capability, and personnel.

Unit – That organization element having functional responsibility for a specific incident's Planning, Logistics, or Finance/Administration activity.

Water Tender – Any ground vehicle capable of transporting specified quantities of water.

B

Integrated Communications

Communications at the incident are managed through the use of a common communications plan and an incident-based communications center established solely for the use of tactical and support resources assigned to the incident. All communications between organizational elements at an incident should be in plain English ("clear text"). No codes should be used, and all communications should be confined only to essential messages. The Communications Unit is responsible for all communications planning at the Incident. This will include incident-established radio networks, on-site telephone, public address, and off-incident telephone/microwave/radio systems.

Radio Networks

Radio networks for large incidents will normally be organized as follows:

Command Net

This net should link together: Incident Command, key staff members, Section Chiefs, Division and Group Supervisors.

Tactical Nets

There may be several tactical nets. They may be established around agencies, departments, geographical areas, or even specific functions. The determination of how nets are set up should be a joint Planning/Operations function. The Communications Unit Leader will develop the plan.

Support Net

A support net will be established primarily to handle status-changing for resources as well as for support requests and certain other non-tactical or command functions.

Ground-to-Air Net

A ground-to-air tactical net may be designated, or regular tactical nets may be used to coordinate ground-to-air traffic.

Air-to-Air Net

Air-to-air nets will normally be predesignated and assigned for use at the incident.

Appendix C
Sample Tactical Worksheets

Sample 1

Tactical Worksheet

Address: ______________________

Occupancy: ______________________

Incident No. ______________________

Time ______________________

Wind Direction

____

Elapsed Time

5 10 15 20 25 30 PAR

Level II Staging

____

Personnel Accountability (PAR)

All Clear

30 Min.

Under Control

Off-To-Def

Hazardous Event

No "PAR" Upgrade Assign.

Tactical **Benchmark** **Functional**

Tactical	Functional
☐ Overall Plan	☐ Command Location
☐ Water Supply	☐ Pumped Water
☐ Search & Rescue	☐ Gas
☐ Initial Attack	☐ Electrical
☐ Exposures	☐ Recon
☐ Rapid Intervention Team	☐ Outside Agency
☐ Logistical Needs	☐ Investigator
☐ Ventilation	☐ P.P.V.
☐ Evacuation	☐ P.D.
☐ *All Clear*	☐ Primary -
☐ *Fire Control*	☐ Secondary
☐ *Salvage (Loss Stopped)*	☐ Salvage (Loss Stopped)
☐ *Accountability*	☐ *C.O. Meter*

E		
E		
E		
E		
L		
L		
H		
R		
U		
BC		

E		
E		
E		
E		
L		
L		
H		
R		
U		
BC		

Branch

Command

Branch

Incident Briefing	1. Incident Name	2. Date Prepared	3. Time Prepared

4. Map Sketch

201	ICS 3-82	Page 1	8. Prepared By (Name and Position)

7. Summary of Current Actions

201	ICS 3-82	Page 2	

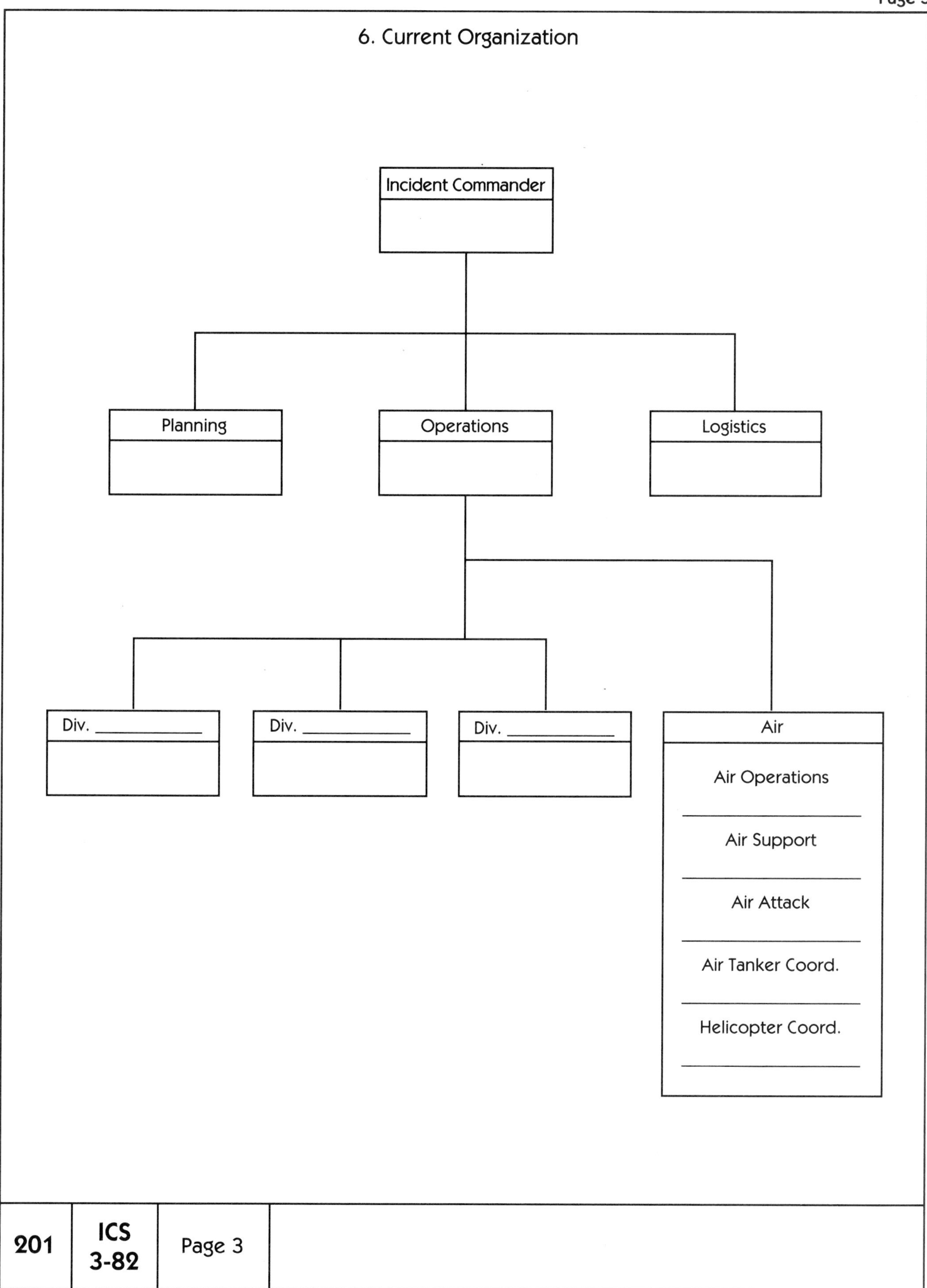
6. Current Organization
Incident Commander
Planning
Operations
Logistics
Div. ____________
Div. ____________
Div. ____________
Air
Air Operations
Air Support
Air Attack
Air Tanker Coord.
Helicopter Coord.

5. Resources Summary

Resources Ordered	Resource Identification	ETA	On Scene ✓	Location/Assignment

Sample 1
Front side

Resource and Situation Status Record

Single Story Structure/ Brush and Grass

Time/Alarm	Address Unit No.		Inc. No.	Date
Time/Scene	Occupancy - Size/Type		Property Loss	
1st Scene	DBA/Occupant	Telephone	Contents Loss	
Agent Appl.	Owner	Telephone	Civ. Inj.	Civ. Fatal.
Knockdown	Owner Address		F.D. Inj.	F.D. Fatal.

Resources			
ENR.	ONS	Assignment	AVI

Assignments	
I.C.	Plans
Operations	Logistics
Lobby	Staging
Base	

	Code 20		Helicopters		Heavy Utility
	C.P. & Base Location		Helicopter Support		Public Utilities
	Police		Tractors		Field Phones
	Emergency Air		Command Unit		Food Service
	Chief Officers		Scat		Weather
	EMS		Mobile Lab		Arson Unit
	Rescue Ambulance		Foam Carrier		
	PIO		Emergency Lighting		

Sample 2
Back side

Resource and Situation Status Record

Multi-Story Structure

Time/Alarm	Address	Unit No.	Inc. No.	Date
Time/Scene	Occupancy - Size/Type		Property Loss	
1st Scene	DBA/Occupant	Telephone	Contents Loss	
Agent Appl.	Owner	Telephone	Civ. Inj.	Civ. Fatal.
Knockdown	Owner Address		F.D. Inj.	F.D. Fatal.

Resources			
ENR.	ONS	Assignment	AVI

Situation	Resources

Assignments	
I.C.	Plans
Operations	Logistics
Lobby	Staging
Base	

	Code 20		Rescue/Evacuation		Salvage
	C.P. & Base Location		Recon		Field Phones
	Police		Helicopters		Public Utilities
	Emergency Air		Ventilation		Food Service
	Chief Officers		Air Conditioning		Arson Unit
	EMS		Stairwells		
	Rescue Ambulance		Elevators		
	PIO		Standpipe Supply		